A Field Guide to Digital Color

# A Field Guide to Digital Color

Maureen C. Stone

CRC Press
Taylor & Francis Group

AN A K PETERS BOOK

First published 2003 by A K Peters, Ltd.

Published 2019 by CRC Press
Taylor & Francis Group
6000 Broken Sound Parkway NW, Suite 300
Boca Raton, FL 33487-2742

ISBN-13: 978-1-56881-161-1 (pbk)

**Visit the Taylor & Francis Web site at**
**http://www.taylorandfrancis.com**

**and the CRC Press Web site at**
**http://www.crcpress.com**

The following are registered trademarks:
Photoshop, PDF, Postscript (Adobe), Cromalin (Dupont), Matchprint (Kodak), and RenderMan (Pixar).

**Library of Congress Cataloging-in-Publication Data**

Stone, Maureen C.
      A field guide to digital color / Maureen C. Stone.
         p. cm.
      Includes bibliographical references and index.
      ISBN 1-56881-161-6
      1. Computer graphics. 2. Image processing--Digital techniques. 3. Color.
4. Coding theory. I. Title.

T385.S7965 2003
621.36'7--dc21
                                                                              2003053579

To My Family

# Contents

# Preface

Color is a part of many different fields, from vision and perception to art and design. The subset of color, called *digital color*, in which color is encoded and manipulated as digital values, still spans many fields, including: image reproduction, computer graphics, and multi-media design. Each field has its own reference books, technical publications and conferences, making it easy to focus on one field without being aware of relevant work in others. This book is a "field guide," designed to provide a high-level summary of each field, with pointers to in-depth resources. For those interested in digital color, this is the book to read first.

## Who Should Read this Book?

- Students learning about color, especially those in computer science, engineering, and digital media.
- Engineers and computer scientists who need to learn about digital color.
- Graphic artists and digital photographers, who are tired of books that present only "how to" without describing "why?"
- Experts in one field of color who want to learn about other fields.
- Anyone wanting a broad introduction to digital color.

# How this Book Is Structured

Each chapter provides a summary of a color field, with a focus on digital color—many fields spread widely beyond pixels. One of the challenges of a book such as this is to balance between superficial and overwhelming. Ideally, experts will find the description of their field an accurate but high-level summary. For those unfamiliar with the field, the description should seem detailed to the point of needing careful reading. Along with providing information specific to the field, the presentation cross-references other fields, and emphasizes common concepts.

Each chapter identifies one or more significant books for its field towards the end of the introduction. Other books may be mentioned throughout the chapter as needed. All books are referenced informally by name, with the full citation information included in the bibliography. Within the chapters, there are occasional references to technical papers that provide detail not found in any book. For these, the full citation information is contained in the margin near where the paper is mentioned. There was a deliberate decision to be selective, rather than inclusive, in the choice of these references, as it is impossible to reference all potentially relevant literature in a book of this breadth without overwhelming the reader. Many of the named papers are good starting points for finding other papers in their area.

There is a resources section at the end of the book that includes a bibliography and pointers to professional organizations, journals, and standards bodies. The list of books is comprehensive, but far from exhaustive, and contains primarily those books I personally own and have found useful. The bibliography is introduced by a discussion that includes a sentence or two about each book. This is not a true annotated bibliography, with a detailed paragraph for each entry, but an effort to qualitatively describe the strengths, weaknesses and application of each book. The description of each organization and standards body includes the journals and magazines published by the organization, and a website for further information.

The book does include mathematical equations to precisely describe concepts such as the transformation between different color representations. Formulas are valuable for implementers, and for those for whom the math speaks more clearly than words. The math is presented, however, in such a way that it can be skimmed or ignored by those not interested in that level of detail.

# Overview of the Chapters

**Chapter 1: Color Vision.** The basic principles of color vision, in which light strikes the eye and is encoded in the retina as exactly three signals, underlie the use of RGB for encoding color. These same principles are used as the basis for measuring color, including representing color as points on the CIE chromaticity diagram, examples of which are used throughout the book.

**Chapter 2: Color Appearance.** The basic principles presented in Chapter 1 provide a simple model for color perception that expands the principles of color appearance. This is best demonstrated by the various optical illusions showing that what a color "looks like" depends on context; for example, in the right circumstances, the identical color (as measured) can appear either black or white. Included is an overview of the computational models that are being used to incorporate these principles into digital color systems. Many color appearance principles underlie color design principles.

**Chapter 3: RGB and Brightness.** In digital color, colors are defined as numerical values, which have no inherent visual meaning. A fundamental requirement for understanding digital color as presented in this book is knowing how to bind these numbers to specific physical and psychophysical metrics. The principles presented in this chapter, even without the mathematical detail, provide a critical foundation for the rest of the book.

**Chapter 4: Color in Nature.** Color in the natural world is created by the interaction of light with matter. This chapter is included to provide a reference for color synthesis in such digital color fields as computer graphics, and to provide a deeper understanding of some of the physical processes that underlie color reproduction technologies such as printing and photography.

Chapters 5–9 are a set of 5 related chapters on the topic of color reproduction. In this book, the color reproduction fields are presented with an emphasis on digital color reproduction and the use of "device-independent color," which is color defined with respect to the psychophysical metrics described in Chapters 1 and 3.

Chapter 5: Color Reproduction. Traditionally, the field of color reproduction is focused on the production of images that are pictures of the natural world. This chapter provides an overview of the general principles of color image reproduction as it is traditionally practiced in photography, television, and printing.

Chapter 6: Image Capture. In image capture, the "natural world" becomes image pixels. The technology of image capture includes cameras and scanners.

Chapter 7: Additive Color Systems. Additive color systems output image pixels as colored light. The technologies of additive color include displays and digital projectors. Originally designed for television, additive color systems are the most familiar physical form for digital color, due to their use in computer displays.

Chapter 8: Subtractive Color Systems. Digital color images must be converted to subtractive color dyes and inks to be printed on paper or projected as slides or movies. This chapter describes the principles and technologies that underlie printing and photography.

Chapter 9: Color Management Systems. Color management systems, which are based on using device-independent color representations for digital color, can be used to create reliable color transformations from one digital medium to another. This chapter provides insight, for example, on how to create prints that are a good reproduction of the pixels displayed on a monitor. It is not a cookbook for any particular system, but a presentation of the basic principles behind all color management systems, including why problems occur and ways to maintain some degree of control.

Chapter 10: Computer Graphics. The field of computer graphics blends the physical models of Chapter 4 with the principles of color reproduction presented in Chapter 5. This chapter focuses on 3D rendering, which is the process that creates images from numerically-defined 3D models.

**Chapter 11: Color Selection and Design.** The principles of color selection and design presented here are primarily those that define the aesthetic use of color as taught in graphic design. These principles are more algorithmic and rule-based than those outside of the field may appreciate. Other color design principles, such as text legibility, are readily defined in terms of the color appearance principles defined in Chapter 2.

**Chapter 12: Color in Information Display.** The effective use of color for information display can be described by Tufte's principles, the primary of which is to "do no harm." This field includes concepts from color perception, color reproduction, and color design that can be applied to visualization, illustration, and user interface design.

# Acknowledgements

My education in digital color began at the Xerox Palo Alto Research Center, where I first tried to print illustrations that had been designed on a computer display. This led to work with Bill Cowan and John Beatty that was one of the first efforts to apply color science to the problem of printing computer-generated images. Bill, who was working with Günter Wyszecki at the National Research Council of Canada, provided expertise in color science and key insights about merging color and digital technology. We received the invaluable help of Dusty Rhodes, who was one of Xerox's resident experts on color printing, with a long and venerable career in the graphic arts. He patiently taught us about high-quality printing and color image reproduction, whose subtleties are not easily modeled in digital color systems. This experience gave me my initial expertise in digital color and my on-going enthusiasm for the study of color.

Throughout my career at PARC, I had opportunity to learn from other color folks at Xerox, including: Gary Starkweather, Mik Lamming, Dale Green, Chuck Haines, and Rob Buckley. Another important part of my professional experience has come from my participation in the SID/IS&T Color Imaging Conferences, where I've had the opportunity to meet and learn from Bob Hunt, Mark Fairchild, Joann Taylor, and Lindsay MacDonald, among many others.

They say it takes a village to raise a child. It can also take a "village" to create a book, especially if one is a working parent. Special thanks goes to Joel Bartlett, who read each chapter as I wrote it and helped me decide when it was "good enough." Joel also helped with technical information about photography and digital cameras. Joann Taylor checked my color science and set me straight about color difference formulae. Bill van Melle proofread the entire book, correcting the grammer and improving the presentation. Bethany Kanui created the index, and Mark Wong helped with the illustrations. Russ Atkinson printed many drafts and lent a sympathetic ear. Jan Hardenbergh offered regular encouragement, feedback, and a helping hand.

Iris, Alice, and the staff at A K Peters provided flexible support, hard work, and a patient tolerance with my perfectionism and resulting delays. They were a pleasure to work with.

Finally, I want to thank my husband, Doug Wyatt, and my sons, Robin and Kevin, for their loving and patient support. Like many books, this one took far more time than expected. Thanks, guys— mom has *finally* finished her book.

## Contributors

There are many people who provided images or data used in this book: Arthur Amezcua, Jason Babcock, Joel Bartlett, Steve Bennet, Cynthia Brewer, Michael Cohen, Elaine Cohen, Mike Cook, Jeffrey DiCarlo, James  Ferwerda, Andrew Glassner, Amy Gooch, Bruce Gooch, Steven Gortler, Donald Greenberg, Radek Grzeszczuk, Chet Haase, Eric Haines, Pat Hanrahan, Jan Hardenbergh, Jack Holm, Paul Hubel, Henrik Wann Jenson, Brad Johanson, Jeremy Johnson, Sterling Johnson, Joe Kane, Russell Kerschmann, Marc Levoy, Bruce Lindbloom, Steve Marschner, John McCann, Lindsay MacDonald, Aditi Majumder, Greg McNutt , Gary Meyer, David Nadeau, Thor Olson, Mary Orr, Sumanta Pattanaik, Arjun Ramamurthy, Erik

Reinhard, David Salesin, Peter Shirley, Michael Stark, Richard Szeliski, Joann Taylor, Kenneth Torrance, Ben Trumbore, Pauline Ts'o, Brian Wandell, and George Winkenbach.

## Special Contributors

Bruce Lindbloom created custom renderings of color gamuts for this book and provided helpful information about the practice of color management in the graphic arts. Bruce's website is a valuable resource for digital color: www.brucelindbloom.com.

Pauline Ts'o scanned in a large selection of her lovely photographs for use in this book, and also taught me about color management at Rhythm & Hues.

Andrew Glassner created several custom illustrations of basic graphics techniques for this book. Andrew's website is: www.glassner.com.

Sterling Johnson sent me an entire CDROM full of stunning bubble pictures. I wish I could have used more of them. Sterling's website is: www.bubblesmith.com.

David Nadeau, from the San Diego Supercomputer Center, told me the fascinating story of how they created the colors for the emmision nebula simulation produced for the Hayden Planetarium in New York City.

Pat Hanrahan and the Stanford University Graphics Lab contributed the use of their color measurement equipment, file server space and the opportunity to learn about projection displays.

# 1
# Color Vision

Color vision starts with light striking the retina, a photosensitive organ in the back of the eye. The cells in the retina encode the light into signals, which are transmitted to the brain. There, the signals are interpreted to provide the perception of color. This chapter focuses on color vision at the encoding level, where it is represented by three signal values generated by the retina. This encoding is the foundation for color measurement, and for the representation of color as a triple of red, green, and blue values.

## Introduction

Light enters the eye as a spectrum of colors distributed by wavelength. The spectrum impinges on the retina in the back of the eye, and for color vision, is absorbed by special cells called *cones*. Human beings have three types of cones, which respond to different wavelengths of light. These are called long, medium, and short

1

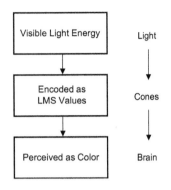

Figure 1.

Process diagram of basic color vision. Light strikes the cones, which encode signals for the brain.

wavelength cones (LMS), or, more informally, red, green, and blue cones (RGB).

Each cone absorbs light and sends a signal along the optic nerve that indicates the total amount of light energy "seen" by the cone. This signal is interpreted in the brain to create the perception of color (Figure 1). This process has two important implications for understanding color vision. First, the spectral distribution is collapsed into exactly three values, one for each cone. Second, these values, not the spectrum, are what represents color at its earliest perception. These principles are called *trichromacy* and *metamerism*, respectively.

Color measurement, or *colorimetry,* can use these principles to encode color as three numbers. Similarly, digital color imaging uses these principles to encode color as RGB pixels, which are also represented by three numbers. This is not coincidence: color vision at the retinal encoding level is essentially an RGB encoding system. Basic vision, as presented in this chapter, is wonderfully orderly, involving concise mathematical operations on spectra and their encodings. However, as the next chapter will show, encoding is only the beginning of color perception. The full perception of color is much less tidy.

There are many excellent books on color vision, several of which are listed in the bibliography. For understanding digital color, my favorite is Brian Wandell's *Foundations of Vision,* which is structured in three parts: encoding, representation, and response. What this chapter describes is essentially the encoding part of his taxonomy. *Color Science,* by Wyszecki and Stiles, is considered the standard reference for quantitative color science and colorimetry. Roy Berns' update of the classic Billmeyer and Saltzman *Principles of Color Technology* is an up-to-date reference on color, its perception, and its measurement.

The first sections in this chapter define the basics of color vision: describing color as a spectrum, the retinal sensors, and the principles of trichromacy and metamerism. The second half describes the principles behind colorimetry, especially those specified by the Commission Internationale de l'Eclairage (CIE).

# Light

Visible light can be described as a function of power versus wavelength. This function is called a *spectral distribution function*, or *spectrum*. Colored light contains wavelengths in the range 370–730 nanometers (nm). The color of the light depends on the distribution of the energy over the visible spectrum; different wavelengths appear different colors.

Each wavelength is associated with a color. Wavelength colors vary from purple (violet) and purple-blue (indigo) through blue, green, yellow, and orange to red, which are the familiar colors of the rainbow. Longer than visible wavelengths are infrared and shorter ones are ultraviolet light. Figure 2 shows the spectral distribution for a fluorescent light bulb. The wavelength colors are roughly illustrated by the colors in the band under the figure. The true colors would be much more intense.

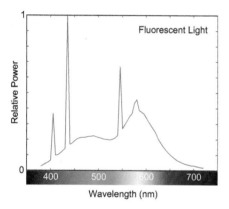

Figure 2.
The spectral distribution function for a fluorescent light. The colored band shows approximate wavelength colors.

The height of a spectral distribution function indicates its power—the taller the brighter. Two spectra that are simply scaled multiples of each other will have the same color, one brighter than the other. For example, Figure 3 is the same purple color at different brightness levels. The area under the spectrum is a measure of brightness called the *intensity*.

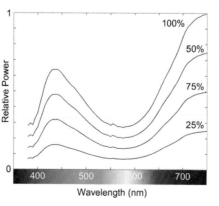

Figure 3.
The same purple color at 100%, 75%, 50%, and 25% intensity.

3

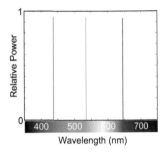

Figure 4.

Three monochromatic colors. Viewed individually, each is a pure, saturated blue, green, or red, as indicated. Viewed all together, the color would look white.

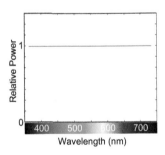

Figure 5.

Equal energy white—the spectrum is the same at all wavelengths.

Spectral distributions that include only a single wavelength describe an intense, pure, saturated color, as in laser light. Such colors are plotted as spikes, as shown in Figure 4 (the colors are approximate).

Spectral distributions that contain all colors in approximately equal amounts will appear white. *Equal energy white* is the color created by presenting equal energy at all wavelengths (Figure 5). While not readily found in nature, it is often used in computer graphics and other computational processes as a canonical white light.

Different mixes of wavelengths will give different colors. For example, the monochromatic colors in Figure 4 are each a pure, saturated color. Viewed all together as a mixture, they will appear white. The precise color seen depends on both the spectrum and the cones, as described below.

Note that it is more common in signal processing and engineering to plot spectra as a function of frequency, which is the inverse of wavelength. This means that red is a low-frequency color and appears on the left of the graph, and blue-violet on the right. This is also the derivation of the terms infrared and ultraviolet. However, color and vision scientists consistently use wavelength plots, where red is on the right.

## Colored Objects

The color of an object is defined by two spectra: the surface reflectance of the object, and the light source shining on it. The product of these two spectra is the light that enters the eye and stimulates the cones. Surface reflectances are often specified as percentages, where 100% reflects all of the available light.

Changing the light creates a different spectral result, resulting in a different color. This is why paint samples in the store, for example, look different than when viewed at home. This is illustrated in Figure 6, which shows the result of multiplying the same spectrum by two different light sources, one representing daylight, and the other incandescent light. In this plot, both surface reflectance

4

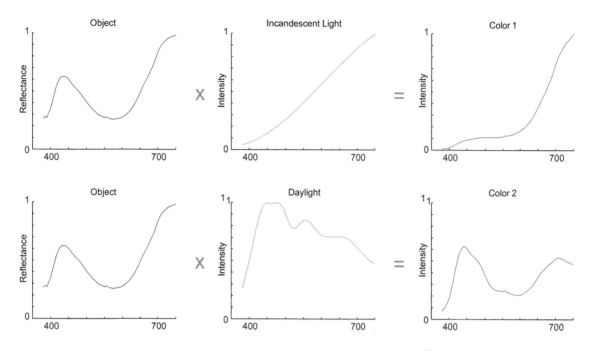

Figure 6.
Different lights reflecting off the same surface create different spectra, and hence different colors.

and intensity are normalized to the range 0...1, to make them easier to evaluate.

The full perception of object color is complex. For example, a painted wall in a room with the light from both a lamp and a window shining on it is perceived to be all one color, even though the light varies significantly across it. Some of these effects are treated in Chapter 2, which describes color appearance. Within this chapter, the color of an object is simply a spectral distribution, which changes as the light changes.

## The Retina, Rods, and Cones

The retina covers the back of the eye. Light enters through the pupil, and is focused by the cornea and the lens on the retina, similar to the way an image is focused on film or on the imaging array in a digital camera. There are two types of imaging sensors on the retina—the *rods* and the *cones*. Only the cones are used for normal color vision.

We see color when the light is bright. As the total light dims (dim room light, or twilight), color vision begins to simplify towards the gray, colorless view one sees "in the dark." Both rods and cones contain photopigments, which generate signals on the optic nerve when exposed to light. Rod photopigments are sensitive at low light levels, but saturate at about the point at which cone photopigments begin to respond. Rod and cone vision are formally called *scotopic* and *photopic* vision, respectively. There is a narrow range of lighting where both rods and cones are active, which is called *mesopic*. Color vision as described here is strictly photopic, involving only the cones.

The response of each cone can be encoded as a function of wavelength called the *spectral response curve* for the cone, as shown in

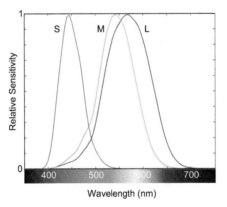

**Figure 7.**

The spectral response curves for the short (blue), medium (green), and long (red) cones.

Figure 7. Note that cone response curves overlap significantly, especially the medium (green) and long (red) curves.

Multiplying the spectrum of the incoming light, called the *stimulus*, by the response curve and integrating to get the resulting intensity defines the signal that is sent from the eye to the brain. This is illustrated in Figure 8. The viewed light is multiplied by the three cone response curves producing three filtered "views" of the spectrum, one for each cone. The area under each of these curves is the signal sent on the optic nerve, as illustrated by the height of the bars.

Unlike manufactured arrays of light sensors, such as those used in digital cameras, which are laid out in a uniform rectangular array, the rods and cones are unevenly distributed in the retina. The cones are concentrated in the center of the eye, in the region called

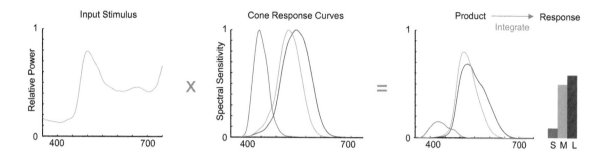

Figure 8.
Multiplying a spectrum by the cone response curves and integrating creates the basic color signal to the brain. The height of the bars reflects the strength of the three signals.

the *fovea*. The rods, plus a few, scattered cones, cover the rest of the retina. There are approximately 5 million cones and 100 million rods in each human eye. There is also a blind spot in the retina, a region that has no rods or cones, where the retinal receptors attach to the optic nerve at the back of the eye. The fovea covers less than 10% of the retina, yet is responsible for all of the color signals sent from the eye to the brain.

How does this uneven mosaic of receptors create a continuous image in our brains? The answer is that our eyes constantly move, in a pattern of rapid jumps called *saccades*. It is left to the brain to stitch together all the images to create a continuous color image.

## Trichromacy and Metamerism

The encoding created by the cones means that every spectrum is represented by exactly three signals—this is the principle of trichromacy. That color can be described by exactly three values is a result that appears over and over again in color science and engineering. Colored pixels in digital color are encoded as three values (red, green, and blue) due to trichromacy.

Different animals have different models for color vision. While most primates are trichromats, birds and fish often have more sensors, indicating that their visual system operates in four or more dimensions. Animal cones may respond to different wavelengths than human cones. Many insects, for example, have sensors that operate in the ultraviolet range, which is invisible to people. Other

7

animals, for example, cats and dogs, appear to have fewer cones than humans, and current thinking suggest that they may see color only in the blue-green part of the spectrum. Previously, it was thought that they saw only shades of gray, though probably with more precision than humans do.

The principle of metamerism states that different spectra that produce the same encoded signals look the same color. Color is defined by the product of the spectrum with the cone response, not the spectrum alone. The principle of metamerism underlies all color reproduction technologies. Instead of reproducing the original spectral distribution that described the color, they create an equivalent response, or *metameric match* by mixing three colors in a controlled

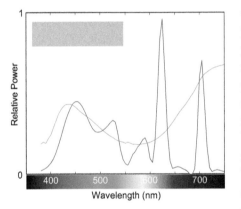

Figure 9.

Two different spectra create the same color, shown as a purple bar, illustrating the property of metamerism.

way. Figure 9 shows a pair of spectra that create (approximately) the same purple color as shown: one is smooth, taken from a colored object lit by white light source, and the other is a mixture of red, green, and blue phosphor emissions from a computer display. These spectral distributions are very different, but appear the same.

Trichromacy and metamerism can be used to create instruments that measure color. Creating an instrument that responds in the same way as the human eye is critical for many industries. It allows colored materials to be described impartially and quantitatively, and creates metrics that define when colors match. From the discussion above, it might seem obvious to fit an instrument with filters and sensors that behave like the cones. However, the precise definition of the cone response was not known until relatively recently. The science of color measurement is much older, and is based on experiments involving matching colors with sets of three primary lights.

# Colorimetry

Colorimetry is the science of color measurement. It is based on empirical studies of humans matching colors. This results are used to create three *color-matching functions* that can be used convert any spectrum into a standard encoding, just as the cones convert any stimulus into LMS signals.

The color-matching experiments that underlie colorimetry are constructed as follows. Choose three primary lights (call them red, green, and blue), which can vary in intensity. Then, take a set of reference colors such as the monochromatic colors of the spectrum, or colors generated by filtering a white light. The job of the observer is to combine and adjust the primary colors to create a result that matches the reference color. This is shown schematically in Figure 10. Once the match is made, the color can then be defined simply by describing the amount of each primary needed to match it.

This is a remarkably useful result. It means that any color can be described as three numbers, which correspond to the amount of each primary used to create the color. These three numbers are called the *tristimulus values* for the color.

What about colors that cannot be matched? The answer is to use "negative light." For example, if the blend is too reddish, even with

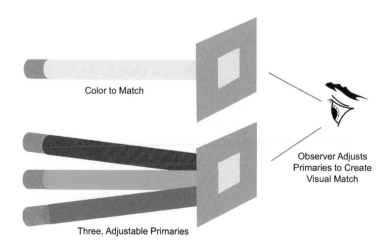

Color to Match

Observer Adjusts
Primaries to Create
Visual Match

Three, Adjustable Primaries

Figure 10.

Color-matching experiments are the basis for colorimetry. An observer adjusts three primary lights to match each sample color.

no red shining on the blend, the observer is allowed to shine some of the red primary on the reference to make the match. The amount of red shone on the reference color is the "negative red" included in the match. It is therefore possible to match all other colors with *any* set of three, distinct colors, assuming you can use "negative light" as described above. In this context, "distinct" means that no primary color is simply a brighter version than another primary.

Defining a color by three numbers plus a known set of primaries is clearly a substantial simplification over the full spectral representation of the color. Color matches defined this way, however, are defined only for colors described as spectrum. For colored objects, the color must be viewed on a standard background under standard lighting to produce a single spectrum as a stimulus. The resulting match, therefore, is valid only for those lighting and viewing conditions.

It would be very inconvenient to have to perform a color-matching experiment to define the tristimulus values for each new color. Fortunately, color matching can be defined more generally because colored light and its perception work as an additive system—the result of shining several lights on the same point is simply the sum of their spectra. A spectrum, therefore, can be considered as the sum of a number of monochromatic colors, one per wavelength. This was alluded to in Figure 4, where the sum of the three monochromatic colors is perceived as white.

Color matching is also additive; the match to a sum of spectra is the sum of the matches. To spell this out mathematically, for a given spectrum, S, let the tristimulus values be R, G, and B (RGB). If $RGB_1$ matches $S_1$, and $RGB_2$ matches $S_2$, then $RGB_1 + RGB_2$ will match $S_1 + S_2$. This principle was first formalized by Grassmann, and it is called Grassmann's additivity law. This principle can be applied to predict the colors produced by additive color systems such as digital color displays, as described in Chapters 3 and 7.

Now, let us return to the problem of color matching. To find the tristimulus values for an arbitrary color, we use Grassmann's additivity law to construct a set of color-matching functions for our primaries. First, we perform the color-matching experiment on each of the monochromatic spectral colors (sampled every nanometer, for example). The result is a set of three functions of wavelength, one for each primary, whose height is the amount of the primary needed to match that particularly wavelength.

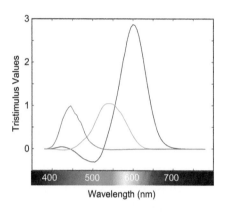

Figure 11 shows a set of color-matching functions constructed by Stiles and Burch in 1959 using a set of three monochromatic colors, as indicated. The red curve is negative in the shorter wavelengths, indicating that some "negative red" is needed to create a match in those regions. All color-matching functions measured from real lights include negative lobes because the long and medium cone response curves overlap. Therefore, there is no way to fully stimulate the red and green cones independently of each other.

Figure 11.

The color-matching functions defined by Stiles and Burch using three monochrome light sources: R = 645 nm, G = 525 nm, and B = 444 nm.

Given a set of color-matching functions and a color defined by a spectrum, the tristimulus values are determined by multiplying the spectrum by the color-matching functions and integrating the results. This is the same process shown in Figure 8 to illustrate the cone response, which suggests that color-matching functions are equivalent to cone response curves. To explore this, we need to look further at the mathematical properties of color mixture.

Given a set of color-matching functions, it is possible to create equivalent functions for a new set of primary colors by linear transformation. All we need to know is the definition of the new primaries in terms of the old ones. For those familiar with linear algebra, this is simply a change of basis, and can accomplished for a three dimensional system with a 3 × 3 transformation matrix. Applying this matrix to each of the samples in the color-matching functions in turn creates the color-matching functions for the new set of primaries. As a result, all color-matching functions are linear transformations of each other.

The cone response functions can be shown to be a linear transformation of the color-matching functions once the optical properties of the eye, such as chromatic aberration, are properly included. This result was finally demonstrated in the late 80s, over 50 years after the first color-matching functions were measured.

# CIE Colorimetry

In 1931, the Commission Internationale de l'Eclairage (CIE) standardized a set of primaries and color-matching functions that are the basis for most color measurement instruments used today. They transformed a set of measured color-matching functions, similar to those of Stiles and Burch, to create a set of curves that were more convenient to use. These are notated $\bar{x}$, $\bar{y}$, and $\bar{z}$ and are shown in Figure 12. This set is positive throughout the entire visible spectrum, and one of the curves ($\bar{y}$) can be used to compute the perceived brightness of the measured color, called its *luminance*. Luminance will be discussed in more detail in Chapter 2. The function z is zēro for most wavelengths because it was more convenient for the hand calculations that were common at the time. There are no physical lights that can be used to create these functions; the primaries are now mathematical abstractions.

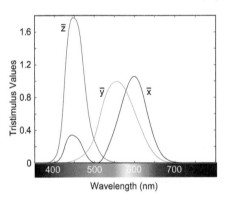

Figure 12.

CIE recommended color-matching functions (1931), also called the CIE 1931 standard observer. $\bar{y}$ can also be used to calculate luminance.

The CIE tristimulus values are computed from the CIE color-matching functions and are notated X, Y, and Z. The color-matching functions have been normalized so that an input spectrum with a constant value of one (equal energy unit spectrum) would create (X, Y, Z ) = (1, 1, 1), as shown in Figure 13. If all input spectra are normalized to lie in the range (0, 1), the resulting tristimulus values

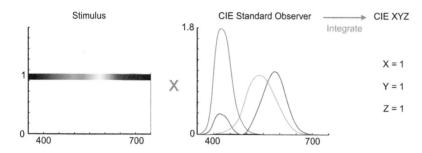

Figure 13.

The CIE tristimulus values are calculated by multiplying the stimulus times the color-matching functions and integrating the result. They are normalized such that equal energy white gives XYZ = (1, 1, 1), as shown.

all fall inside the unit cube—no tristimulus value would exceed 1.0. Both absolute and normalized tristimulus values are used in color measurement.

It is difficult to visualize and compare three-dimensional values, and in many applications, it is useful to be able to define a color independently from its brightness. The CIE *chromaticity diagram,* shown in Figure 14, is a two-dimensional projection of the tristimulus values that is the most common way of visualizing them. It is derived from the CIE XYZ values by dividing them by their sum, as shown in the equations. The resulting chromaticity values, notated x, y and z, have the property that x + y + z = 1. The chromaticity diagram is a plot of x versus y.

This diagram is worth some study. Any color (a spectrum of light) can be plotted as a point, independent of its brightness (power). Also, all metameric matches to the spectrum, which by definition have the same tristimulus values, will plot at the same point.

The chromaticity diagram illustrates how color is additive. Take two colors, each of which plots as a point in the chromaticity diagram. Any colors that are mixtures of these two colors will plot along the line connecting them. The distance along this line is proportional to the relative brightness of the two colors. For three colors, all mixtures will lie within the triangle formed by their chromaticity coordinates, and so on.

The monochromatic spectral colors lie along the horse-shoe shaped path, which is called the *spectrum locus.* All visible colors lie within the shape bounded by this path and the line that connects

Figure 14.

CIE 1931 chromaticity diagram showing the spectrum locus, the purple line, and the black-body curve. The transformation from CIE XYZ is shown in the equations.

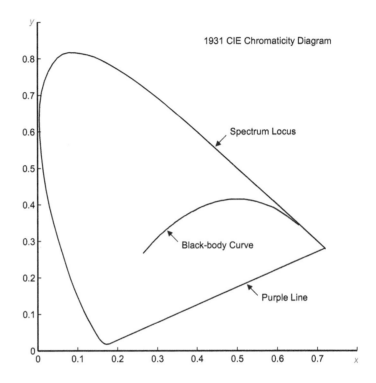

1931 CIE Chromaticity Diagram

Spectrum Locus

Black-body Curve

Purple Line

$$x = \frac{X}{X + Y + Z}$$

$$y = \frac{Y}{X + Y + Z}$$

$$1 = x + y + z$$

the two ends (called the *purple line*) because all spectra can be created by sums of monochromatic colors. The closer a color lies to the spectrum locus, the more saturated (vivid, colorful) it is. Take a point near the center of the diagram as a reference white and draw a line out to the spectrum locus. The line defines a family of colors that have the same basic hue, but become progressively more pure (saturated) as they approach the spectrum locus. The colorimetric term *purity* describes the relative distance along such a line.

Colors that appear white lie near the center of the chromaticity diagram. Most of these colors are clustered near the *black-body curve*, which describes colors caused by heating an object that reflects no light (called a black-body radiator) until it glows white-hot. These colors range from yellowish-red to blue-white.

14

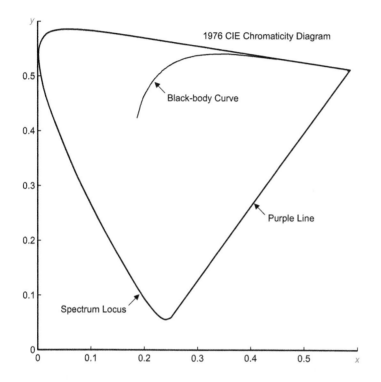

1976 CIE Chromaticity Diagram

Black-body Curve

Purple Line

Spectrum Locus

Figure 15.
CIE 1976 chromaticity diagram showing the spectrum locus, the purple
line, and the black-body curve. The
transformation from CIE XYZ is shown
in the equations.

$$u' = \frac{4X}{X + 15Y + 3Z}$$

$$v' = \frac{9Y}{X + 15Y + 3Z}$$

One major concern with the 1931 chromaticity diagram is that distances are not proportional to perceptual differences in color. For example, two points that are within, say, 0.1 units of each other in the reddish (lower right) part of the chart will be perceptually more different than two colors the same distance away in the greenish (upper left) part of the chart. Therefore, there have been many variations on this diagram that attempt to make it more uniform. One of the most commonly used is the u′ v′ system, shown in Figure 15, along with its supporting mathematical transformation. Which chromaticity diagram to use is often very much a matter of practice and technical "taste." In spite of the many improvements proposed and adopted throughout the years, the 1931 diagram is still the most common.

# Colorimetry and Colored Objects

Colorimetry can only be applied directly to the light reflecting from a colored object, which can be described as a spectral distribution. Colorimetry applied to objects, therefore, will always depend on the spectral description of the light. Many instruments for measuring colored objects contain a light source, and are held against the surface of the object to eliminate all other light.

Some authors multiply the surface reflectance by the color-matching functions to define "tristimulus values" for the surface. This is a rather suspect practice that stretches the definition of colorimetry beyond its origins in visual matching. Physically, this is equivalent to illuminating the surface with a uniform white light (equal energy at all wavelengths) to create a spectral stimulus. The result, however, is different than any color the object will produce in practice, because such light sources do not exist in nature. Furthermore, such tristimulus values do not follow the additive principles described above. What does it mean to blend two object colors? The potential for confusion hardly seems to compensate for the possible convenience.

# Color Measurement

Instrumentation for tristimulus-based color measurement comes in two forms: instruments that sample spectra and compute tristimulus values based on the samples, and instruments that use filters to approximate the color-matching functions. The first usually have "spectro-" in their name, and are either *spectroradiometers* (which measure light directly and absolutely) or *spectrophotometers* (which measure relative reflectance, and often contain their own light source). The second form are *colorimeters*, and are often specifically designed for monitors, reflection prints, or transmission films. In general, colorimeters are cheaper, faster, and less accurate than systems that sample spectra.

All color instrumentation comes with a specification that tells its precision and the range of brightness it can accurately measure. It is important to pay attention to this information: there is no point in agonizing over a 1% luminance variation, for example, if the instrument is only accurate to 2%. Similarly, measuring something that is too bright may simply generate bad data, not an error message, and may also damage the instrument.

All instrumentation needs routine calibration and evaluation to be sure it is working correctly. In addition to following the manufacturer's recommendations, it is useful to create some standard measurements, and make them routinely to see if the instrument has drifted. Similarly, for any task, it is important to find some "reality check" measurements to ensure you don't waste time gathering garbage. For example, when measuring a CRT display, the sum of the tristimulus values for red, green, and blue should equal the tristimulus values for white, because white is the sum of red, green, and blue.

Most color measurement equipment can be connected to a computer via a serial port. The vendors often offer both a way to interface the instrument to a program and some end-user software for color measurement. My experience is that this software is often not as good as the equipment it runs—these folks understand color measurement, not software. Similarly, be prepared for some quirks in the serial port driver, as the "standards" in this area seem to be rather loosely interpreted, a problem that is not unique to color measurement equipment. It is most convenient if the interface can be driven by sending text codes down the serial port line. This is generally robust, and can be easily interfaced to any system, or even driven by hand from a terminal emulator.

Before color management systems created a need for inexpensive and convenient color measurement, most colorimetric instruments were expensive and difficult to use. The use of color measurement in color management systems (described in Chapter 9) for the graphic arts and digital photography has created a market for less expensive and easier to use equipment. While often not as precise as the high-end instruments, such equipment is usually adequate for digital color imaging needs.

# Densitometric Measurement

*Densitometry* is such a common form of color measurement that it is described here, even though it is not based on color-matching functions and the principles of CIE colorimetry. A *densitometer* measures the amount of light transmitted through or reflected from a material and encodes it logarithmically. *Density* is defined as $-\log_{10}(T)$ or $-\log_{10}(R)$, where T (transmittance) or R (reflectance) lie in the range (0,1). Therefore, a transmittance of 1 is a density of 0, a transmittance of 0.001 (0.1%) is a density of 3.0, and the density for a transmittance of 0 is undefined.

*Color densitometers* measure density through narrow-band color filters, such as those shown in Figure 16. These are designed to selectively measure colorants in photographic or print media. Their spectral response functions are not color-matching functions, and cannot be used to compute tristimulus values. True color-matching functions always overlap, as do the curves in Figure 11 and Figure 12.

Densitometers are often used in digital color to measure the dyes and inks used in print and film, the subtractive color processes. The peaks of their color filters are designed to match the specific colorants being measured.

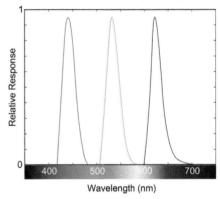

Figure 16.

Color filters for densitometry (ISO Status A), recommended for measuring the colors in photographic prints. These are not color-matching functions.

# Summary

Light in the visible spectrum stimulates the cones in the retina, which encode and send messages along the optic nerve to produce the sensation of color. Different spectra can produce the same response, which is the property called metamerism. The encoding can be represented by three numbers because the human visual system is basically trichromatic (three-colored). The principle of trichromacy is fundamental to color measurement and color reproduction.

Color measurement uses these properties to create representations of color that can be encoded by instruments. These are based on the CIE recommendations of a standard observer (CIE color-matching functions), tristimulus values, and chromaticity diagram. Such measurements provide a metric for specifying colors in many scientific and industrial domains.

Trichromatic color matching can be described by simple, linear mathematics, which makes it very attractive to scientists and engineers. However, it is important to understand its limitations. The tristimulus values describe a spectral stimulus, created either by directly viewing a light or by viewing light reflected from an object. Practically speaking, trichromatic color matching only "works" for a single color on a simple background under controlled viewing conditions. Colors in the real world, or even colored pixels in images, are influenced strongly by their surrounding environment. The appearance of these colors is a far more complex topic, which will be covered in the next chapter.

# 2
# Color Appearance

The models used in colorimetry are remarkably simple. Color is encoded with respect to a three-dimensional, linear model that can be measured with appropriately designed instruments. However, the truth about the perception of color is much more complex. Color matching as described above only works for individual colors viewed under simple, controlled circumstances. Its definition is based on the perception of colored light. In the real world, we see complex arrangements of colors. These colors interact in ways that defy the simple model presented by colorimetry. This chapter will describe the key concepts and models used in the study of color appearance. These influence the application of digital color, both for image and information display.

## Introduction

The trichromatic encoding for color presented in Chapter 1 provides a way to measure colors represented as spectral distribution. For two colors to "look the same," however, they must be viewed alone, without any distracting background colors. Without this constraint, colors that measure the same can appear quite different. Figure 1 is a classic photograph taken by John McCann. In it, two standard color charts are shown, one in the sun and one in the shade. The difference in lighting is such that the measured intensity of the black patch in the sun was exactly the same as the white patch in the shade. That is, two instruments would assign them the same color. However, they clearly look black and white in the context of this photograph. That they are the same intensity can be seen in the image on the right, where the patches have been masked to remove the context. Both patches appear medium gray, and should be identical except for variation introduced by the color reproduction process that produced this book. A similar effect can be seen in the gray step wedge, which lies across the shadow boundary with the dark end in the sun. While it appears to show a monotonically decreasing set of gray steps, black and white again have the same brightness.

The appearance of colors in the real world differs from the simple models defined by the retina due to a combination of perceptual and cognitive effects. The perceptual effects are created by the encoding and processing of the original retinal signals, which is more complex than the simple RGB encoding described so far. Cognitive effects are based on our knowledge of how objects and lights behave in the world.

The field of color appearance tries to quantify the perceptual, and to a lesser extent, the cognitive effects that define color appearance. The goal is to create a more flexible and robust definition of "looks the same" than that provided by trichromatic theory.

Mark Fairchild's book, *Color Appearance Models*, includes both a good description of color appearance as it is currently understood, and a survey of computation models that are being developed to address this complex topic. While the book is now five years old, it is still the best general reference for this topic.

This chapter begins with a description of perceptually defined color spaces. These spaces take a distinct form defined with respect to lightness, hue, and colorfulness, rather than the RGB specification defined at the retina. This form has a psychophysical basis; the

encoding of the cone signals into opponent color channels, which occurs early in the processing of the visual signal.

The second part of this chapter discusses factors that affect color appearance. Color appearance has a strong spatial component that includes the interaction between adjacent colors, especially if one surrounds the other. Size is another strong component of appearance; the smaller the sample, the less saturated it appears. Adaptation, which describes the way the eye and its receptors adjust to different lighting conditions, is another factor, and one that is very important to reproducing colors across different media. Color appearance models attempt to quantify these effects to predict the appearance of colors based on measurable quantities like CIE tristimulus values.

**Figure 1.**
Two identical color charts photographed on a sunny deck demonstrating that the same color can appear either black or white, depending on the context. (Image courtesy of John McCann.)

## Perceptual Color Spaces

If you ask people to describe a color, they do not normally speak in terms of its red, green, and blue components. They will typically name a hue, such as red, purple, orange, or pink. They may describe it as being light or dark, and they may call it vivid, colorful, muted, or grayish. These terms are used to create a perceptual organization of color, as shown in Figure 2.

A perceptually organized color space has these axes, and ideally, the additional property that the space is perceptually uniform: equal

Figure 2.
Figure 2.
The perceptual organization of color, structured as hue, lightness, and colorfulness.

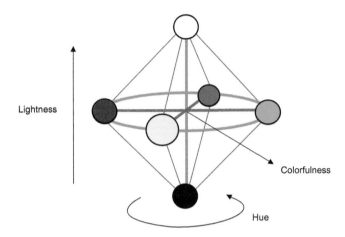

distances define equal perceptual differences everywhere in the space. An example is the Munsell color-order system, which describes each color by its hue, value (lightness), and chroma (colorfulness). The numeric scales assigned to these metrics create a perceptually uniform color space. The original Munsell definition was created with hand-painted color chips, which were visually evaluated for accuracy. Figure 3 is a photograph of a set of Munsell color chips displayed on a Munsell Color Tree, which nicely illustrates the shape and organization of the Munsell color-order system. The Color Tree consists of a number of chroma-value slices through the color space. Each slice shows all the chips of a specific hue with value increasing vertically and chroma radially. Different hues have different sets of colors visible on the slice, showing the extent of the color space, which is bounded by the original specification. In theory, the notation could describe a sphere of colors.

You can still buy *The Munsell Book of Color* or the pictured *Munsell Color Tree* from the Munsell Color division of GretagMacbeth. There are more samples in the book version than shown on the Color Tree. The Munsell chip set has also been defined with respect to CIE tristimulus values, which makes it possible to algorithmically transform between the two specifications. Software to simulate the Munsell colors is available for free download from the Munsell website (www.munsell.com).

CIELAB and CIELUV are non-linear transformations of the CIE tristimulus values that create perceptually organized, (somewhat) perceptually uniform color spaces. These spaces were defined to simplify the computation of color differences. Because they are perceptually uniform, color differences can be specified as Euclidian distances in these spaces; that is, the length of the line between two points is a measure of the color difference. A unit step (length = 1), notated $\Delta E$, describes a "just noticeable difference" or JND. These spaces were designed for evaluating small color differences (less than 5 $\Delta E$), as these are the ones most important for commerce. Therefore, a difference of 20 $\Delta E$, for example, in one part of the color space, may not be better, or significantly different, than a difference of 21 $\Delta E$ elsewhere in the space.

CIELAB and CIELUV have the same lightness axis (notated $L^*$) but compute their other two components differently. The reason there are two spaces is an artifact of the standards process. At the time the CIE was evaluating color difference spaces, there were two competing proposals. Ultimately, the CIE decided they were equally valid and formalized them both. CIELUV has been more commonly

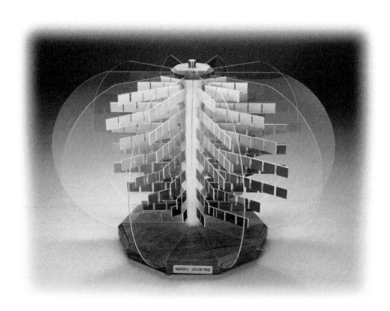

Figure 3.
The Munsell Color Tree, which displays color chips illustrating the Munsell color-order system. Each section is a chroma-value slice for a specific hue. (Courtesy of Gretag Macbeth.)

used in the television and display industries, CIELAB in the printing and materials industries. Over the years, CIELAB has become recognized as a the more accurate of the two. CIELAB has also been used as the basis for work in color appearance spaces, as is described at the end of this chapter.

Figure 4 is a summary of the mathematics of CIELAB. Put into words, CIELAB is a non-linear function of CIE tristimulus values. To create a single lightness axis, its calculation requires a reference white, also specified as CIE tristimulus values ( $X_n$, $Y_n$, $Z_n$,). As a result, white is always L*, a*, b* = (100, 0, 0), and black is (0, 0, 0).

For each component of CIELAB, the first step is to divide the tristimulus values by the reference white, and the second is to compute the cube root. The cube root function is replaced by a linear one, scaled and offset so that the slopes will match whenever the

---

**Equations for CIE 1976 L*, a*, b* (CIELAB)**

$$L^* = 116 \left[ \left( \frac{Y}{Y_n} \right)^{1/3} - \frac{16}{116} \right]$$

$$a^* = 500 \left[ \left( \frac{X}{X_n} \right)^{1/3} - \left( \frac{Y}{Y_n} \right)^{1/3} \right]$$

$$b^* = 200 \left[ \left( \frac{Y}{Y_n} \right)^{1/3} - \left( \frac{Z}{Z_n} \right)^{1/3} \right]$$

$X_n, Y_n, Z_n$ are the tristimulus values of the reference white.

If $\frac{V}{V_n} \leq 0.008856$, where $V$ is any of $X$, $Y$, or $Z$, replace

$\left( \frac{V}{V_n} \right)^{1/3}$ with $\left[ 7.787 \left( \frac{V}{V_n} \right) + \frac{16}{116} \right]$ in the equations above.

**Equations for Hue ($h_{ab}$) and Chroma ($C_{ab}^*$)**

$$h_{ab} = \arctan \left( \frac{b^*}{a^*} \right) \qquad C_{ab}^* = \left[ a^{*2} + b^{*2} \right]^{1/2}$$

Figure 4.

The mathematics of CIELAB, including the basic equations plus the specification for hue and chroma.

ratio of tristimulus values becomes very small. This is to avoid the steep slope of the cube root function near zero. The various scale factors are to make the mathematics fit the experimental data used to derive it. CIELAB, like most formula in color psychophysics, is an empirically derived function.

While $L^*$ has a clear, perceptual meaning (lightness), $a^*$ and $b^*$ do not. Converting to cylindrical coordinates creates a convenient perceptual description. $L^*$ remains lightness, the angular displacement defines hue ($h_{ab}$), and the radial distance defines chroma ($C_{ab}^*$).

There are several ways to compute the difference between two colors in CIELAB. The Euclidian distance between two colors is:

$$\Delta E_{ab}^* = \left[(\Delta L^*)^2 + (\Delta a^*)^2 + (\Delta b^*)^2\right]^{1/2}$$

In some applications, it is more informative to consider the difference between individual components, such as $\Delta L^*$ or $\Delta C_{ab}^*$. Differences in hue can be considered in two ways: the difference in hue angle ($\Delta h_{ab}$) or as the Euclidian distance $\Delta H_{ab}^*$, which is defined as:

$$\Delta H_{ab}^* = \left[(\Delta E_{ab})^2 - (\Delta L^*)^2 - (\Delta C_{ab}^*)^2\right]^{1/2}$$

Mathematically, this is the distance between two colors once the difference in lightness and chroma have been subtracted. As a result, $\Delta E_{ab}^*$ can be described in terms of $L^*$, $C_{ab}^*$, and $H_{ab}^*$ as:

$$\Delta E_{ab}^* = \left[(\Delta L^*)^2 + (\Delta C_{ab}^*)^2 + (\Delta H_{ab}^*)^2\right]^{1/2}$$

This form of the color difference equation is used in newer color difference formulae, such as CIE94. These models add numerical weights to the terms of this equation to compensate for non-uniformities in the CIELAB color space and to fine-tune the difference equation for specific applications. More detail about CIE94, as well as other color difference metrics, can be found in Bern's *Principles of Color Technology*.

In spite of its mathematical description, which can be applied to any color, it is important to understand that the original CIELAB specification was created by experiments comparing painted color chips. That is, its uniformity was only verified for object colors, which form a subset of the full range of visible colors. Colors rendered on a display may fall outside of this region, especially fully saturated colors on monitors. Many of the problems discovered applying CIELAB to digital color imaging applications involve such colors.

# The Opponent Color Model

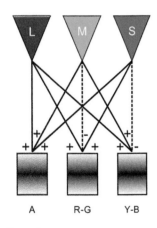

Figure 5.
The LMS cone response signals are combined to create the opponent channels. All cones contribute to all opponent channels in proportion to their density in the retina.

The perceptual organization of color reflects the first level of processing applied to the cone response signals (LMS) and is called the *opponent color model*. In this model, the LMS values from the retina are encoded as achromatic brightness (A) plus a red-green (R–G) and a yellow-blue (Y–B) color difference signal, as shown in Figure 5. As the figure indicates, $A = (L + M + S)$, $R-G = (L - M + S)$, and $Y-B = (L + M - S)$. Each of the L, M, and S values must be normalized to the density of cones of that type, which is surprisingly varied. The relative populations of L:M:S cones are approximately 40:20:1. The scarcity of S cones, which are absent entirely from the center of the fovea, limits the sharpness of intensely blue colors, as will be discussed in Chapter 11.

The difference encoding of color can be seen in the afterimages produced by staring at brightly colored patches. Stare steadily at the dot in Figure 6 for about a minute, then look at a white surface. You will see an afterimage caused by saturating the red-green, blue-yellow, and achromatic opponent channels. In the afterimage, the red square will appear green, the blue one yellow, etc.

The opponent color model also well describes the common forms of color vision deficiencies. While these are usually caused because one type of cone in the retina is defective, they are most easily described in opponent terms. The most common problems are anomalies in the red-green opponent channel, which affects the ability to see red and green. This type of deficiency appears in approximately 10% of men. A much smaller percentage (1–2%) exhibit weaknesses in the blue-yellow channel, with very few people actually "color blind," or unable to see any hues at all. While most color vision problems are genetic, they can also appear as a side-effect of medication or illness. The popular drug Viagra, for example, lists difficulty distinguishing blue from green as a possible side effect.

The effect of color vision deficiencies is to make it difficult to discriminate between colors that are normally distinct, an important factor to consider in the use of color for information display. Often, shape and positional cues can be used to help compensate for this problem, and a good designer will include them. This topic is discussed further in Chapter 12.

One of the most difficult viewing situations for those with color vision deficiencies are intensely colored lights that stimulate primarily the abnormal cone, such as the red LEDs that used to be popular in instrumentation displays. It is interesting that red and green stoplights are not more of a safety hazard, given that 10% of the male population finds them difficult to discriminate by color alone. Fortunately, there are usually positional cues that help identify which signal is which.

Opponent color theory was first proposed by Herring in 1878, but was not accepted until Dorthea Jameson and Leo Hurvich performed controlled experiments in 1955 that verified the theory. It is now accepted as the first-level processing of the color vision signal after the cone response. Leo Hurvich has written a textbook, *Color Vision,* that is based on this theory. It is also the best technical reference I have found for the modalities of color vision deficiencies. It is unfortunately out-of-print, but may be found in libraries.

Figure 6.
Stare at the black dot for about a minute, then look at a white surface to see the afterimages.

# Lightness Scales

Lightness, brightness, intensity, luminance and L* are all ways of arranging a set of colors along a scale from dark to light. As the control of this mapping is key to many color imaging applications, it is important to understand what these quantities mean and how they relate to the perception of color.

Lightness and brightness are qualitative terms, with no specific measurement associated with them. Color scientists when speaking

Reference.
Dorthea Jameson and Leo M. Hurvich. "Some Quantitative Aspects of an Opponent-Color Theory: I. Chromatic Responses and Spectral Saturation." *Journal of the Optical Society of America* 45 (1955), 546–552.

formally, treat brightness as an absolute term associated with emissive sources such as lights. Lightness is a relative term, typically associated with reflective surfaces, but can also mean the brightness with respect to a defined white for an emissive display. Informally, they are often used interchangeably.

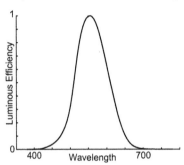

Figure 7.

The luminous efficiency function describes the brightness sensitivity of the eye as a function of wavelength. This is the response for color vision, created by the cones.

*Intensity* is a measured quantity whose units are light power. A typical spectral distribution function describes the intensity at each wavelength. The total intensity of the spectrum is the integral of the function, or the sum of all the wavelength powers. *Luminance* is intensity measured relative to the eye's spectral response. The curve shown in Figure 7 is the *luminous efficiency function*, which describes the relative sensitivity of the eye for each wavelength as defined by the cones (there is a different curve for rods). It peaks in the yellow-green part of the spectrum at a wavelength of 555 nanometers, and tapers to zero at both ends. Take any spectrum, multiply it by this curve, and integrate to get the luminance, which is a measure of the perceived intensity. To better understand the difference between intensity and luminance, consider two lights of equal intensity but of different colors, one green and one blue. The green light will have a higher luminance than the blue one; it will appear brighter for the same intensity because the eye is more sensitive to its wavelengths. Double the intensity of both lights (that is, uniformly scale the spectra) and the luminance will also double. For any given spectrum, there is a scalar relationship between intensity and luminance.

Luminance defines a power scale for perceived color. It can be defined in absolute units, typically candelas/meter$^2$. But for many applications, only relative luminance is importance, so the luminance measurements will be normalized with respect to some maximum value, typically the value for white. The CIE color-matching functions are defined such that the tristimulus value Y is the same as luminance.

Plotting equal steps of luminance does not create a perceptually uniform lightness scale where each step is equally different from the other. One good way to define such a scale is to ask an observer to find the gray color half way between two samples, one black and one white. Then split the difference between the gray and the black, one white. Then split the difference between the gray and the black,

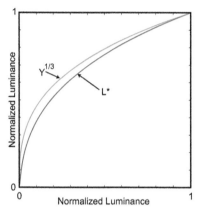

similarly the gray and the white. Repeat until the patches are just noticeably different from each other. The resulting scale will be perceptually uniform.

This process is easy to implement, and people are quite consistent about which values they pick. Another consistency is that they normally pick around 100 patches for the completed scale. This is how the Munsell value scale was defined, for example, using a set of painted samples. The quantity L*, which is the lightness axis of both the CIELAB and CIELUV color spaces, is a function of relative luminance that closely matches the Munsell value scale. Its mathematics were given in Figure 4, and comprise a cube root function plus a linear portion, scaled and translated to fit smoothly together. The resulting function is shown in Figure 8, which is a plot of L* as a function of relative luminance. A plot of relative luminance (Y) to the 1/3 power is included for comparison, showing that the translation and scaling included in the L* equation make a significant difference in the shape of the curve. Fitting this curve with a power function gives an effective exponent of 1/2.43, not 1/3.

There is no single definition of a perceptually uniform lightness scale. The size of the samples, the background they are displayed against, and the overall viewing conditions all affect the results. For example, Figure 9 shows how three identical gray scales with different backgrounds vary in perceived spacing. Note, for example, how your perception of the location of 50% gray moves as a function of background.

Figure 8.

The L* function, contrasted with a cube root function of luminance. The best fit to L* is $Y^{1/2.43}$.

Figure 9.

The same gray ramp is displayed on three different backgrounds, which changes the apparent spacing of the colors.

The lightness of objects is described in terms of their reflectance or transmittance, which indicates the amount of incident light reflected from or transmitted through the object. These quantities are most precisely described as spectral distributions, but it can be useful to describe the total reflectance or transmittance as a percentage, especially when comparing object colors with similar spectral characteristics. For example, a set of gray filters of varying degrees of darkness could be described as having transmittance values of 25%, 50%, and 75%, but this is only precise if the transmittance spectra are scaled multiples of each other.

In color imaging applications, the mapping from pixel values to perceived lightness or brightness is a key part of any digital color specification. Small alterations in this mapping can make large changes in color image appearance. Controlling this mapping through the image reproduction process is one of the key components that affect image quality.

Figure 10.

The same gray square on four different backgrounds, showing the effect of simultaneous contrast. It should look lighter on the black background than the white one; greenish on the red back ground; and reddish on the green one.

Figure 11.

On the bars, all the patches have identical surrounding colors (half yellow and half purple) but appear as if they were completely surrounded by the color of the bar they are "on."

# Spatial Phenomena

Trichromatic theory applies to colors in isolation. Samples to be compared must be a similar size, and viewed on a neutral background. Many color appearance phenomena are caused by the interaction between colors. Absolute size matters also, which is why the paint chip is always less vivid than the painted wall. This section demonstrates some of the spatial phenomena that affect color appearance.

*Simultaneous contrast* describes the influence of immediately surrounding colors on the perception of a color. The simplest model for simultaneous contrast is that the afterimage of the surrounding color is added into the perception of the surrounded color. Therefore, a gray patch on a dark background looks lighter (add white) than a gray patch on a white background (add black), and a gray patch on a

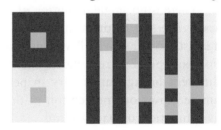

 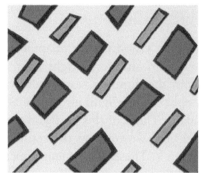

Figure 12.

Bezold Effect: the colors in the patterns are identical, except for the black or white outlines.

red background looks more greenish than a gray patch on a green background (Figure 10). However, the real answer is more complex. Figure 11 shows identical blue-green patches on a purple and yellow background. The color looks quite different. Placing these same colored patches on alternating purple and yellow bars produces the same difference in appearance, even though the actual surrounding is identical for each (half yellow and half purple). It seems that a cognitive effect, in which a patch appears to be "on" a particular colored bar, is contributing to the phenomenon of simultaneous contrast in this example.

Even a thin outline of a different color can make a big difference in the appearance of a colored region. The only difference between the two patterns in Figure 12 is the color of the outlines—white in one case and black in the other, but the overall difference in appearance is quite dramatic. This effect is called the Bezold Effect, named for the nineteenth century rug designer who found he could reuse his designs by this simple alteration.

Size, or *spatial frequency*, has one of the strongest impacts on the perception of a color. The higher the spatial frequency, the less saturated the color. In the bands of alternating bars shown in Figure 13, exactly the same two colors are used. The difference in appearance is entirely due to the difference in spatial frequency.

Figure 13.

The colors in the two patterns are identical. Only the spatial frequency is different, which changes their appearance.

# Adaptation

Human vision is very adaptable. We are capable of seeing in both very dim and very bright light, and over six orders of magnitude from the dimmest starlight to full sunlight. When we move from bright to dim lighting, or vice versa, we can feel our visual system adapt. Think of this as a gain control on the visual system. For dim lighting, we turn up the sensitivity. For bright lights, we need to damp it. These phenomena are called *dark adaptation* and *light adaptation,* respectively. These adaptation phenomena are controlled both by changing the pupil size, which happens fairly quickly, and by changing the sensitivities of the photopigments in the cones and rods. Light adaptation takes only a few tenths of a second, while full dark adaptation can take up to 40 minutes

In many cases, it is valid to assume that color appearance is relative to the current black/white levels, and is not effected by the absolute luminance, as long as the luminance levels are high enough to engage the cones and saturate the rods (photopic vision). However, increasing luminance can produce a measurable increase in colorfulness (called the "Hunt Effect") and contrast (called the "Stevens Effect"). Basic color imaging models do not include these effects, but more sophisticated color appearance models do.

*Chromatic adaptation* describes the visual system's ability to adapt to the color of the light illuminating the scene. Most color is created by light reflecting off of objects. While the reflected spectrum can be measured with colorimetric instruments, changing the light will change the measured color, sometimes dramatically. But, as we view the world, we don't generally perceive objects changing color as the light shifts: There is an automatic "white-balancing" function in the visual system, in which the "gain controls" for the three cones are adjusted separately to compensate for the change in lighting. Modeling chromatic adaptation is very important for the accurate reproduction of digital images.

Figure 14 gives some sense of how strongly color is perceived with respect to the reference white. The original scene, Figure 14(a), shows a boy in a yellow shirt. Figure 14(b) shows the same scene with a

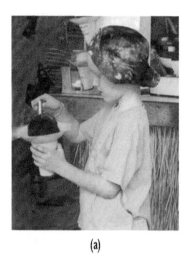  

(a)                                    (b)                                    (c)

strong purple tint applied. The shirt still appears yellow, however, in the context of the picture because all of the colors have shifted together in a way that mimics chromatic adaptation. Figure 14(c) shows the same tint applied only to the shirt, which now appears distinctly pink or peach, but not at all the same yellow as in Figure 14(b), even though the color is exactly the same. This illustration was inspired by a classic example created by Robert Hunt that involved applying cyan filters to a picture of a woman with a yellow pillow.

The cone response physically adapt to the color of the light source. This general concept was first proposed by von Kries in 1902, with the admonition that it was probably too simple to be correct. However, it has proven to be a robust model for chromatic adaptation. Modern interpretations of von Kries depend on the ability to define the cone response functions, and are expressed in terms of scalar values that are the inverse of the maximum response for each cone, as will be described later in this chapter.

Figure 14.

(a) Original image; (b) overall purple tint, but the shirt still looks yellow; (c) tint applied just to the shirt. The shirt color is identical in (b) and (c), but appears different due to adaptation.

# Color Appearance Models

A color appearance model converts from measured values such as the tristimulus values of the sample and its surrounding color, the overall illumination level, etc., to correlates of perceptual attributes of color. The goal is to model the way colors change in appearance as a function of the color of the lighting, surrounding colors, and the overall brightness of the viewing environment (dim, bright, etc.). A driving application for color appearance modeling is digital color reproduction, where the goal is to effectively transform, for example, colors viewed on a CRT in a dim environment, to colors printed on a page viewed in a bright one. Color appearance models have been an active topic in the digital color research community for the last decade, and there is yet more to solve. However, models do exist that provide more accurate appearance predictions than colorimetry, or even the use of color difference spaces like CIELAB.

To model color appearance, we must formally define its attributes. These are described as *hue, chroma, colorfulness, saturation, lightness, and brightness*. These can be associated with the three dimensions of the perceptual color space shown in Figure 2.

Hue describes the attribute associated with the color name, such as red, green, or blue. Hue can be defined in terms of the primary opponent colors: red, green, blue, and yellow, as shown on the hue circle in Figure 2.

Brightness is an absolute term, whereas lightness is a relative term, defined with respect to some reference white. Brightness is usually applied to lights, whereas lightness is applied to objects. Brightness and lightness are terms associated with the lightness axis of Figure 2.

Colorfulness, which is shown as the radial dimension in Figure 2, is also an absolute term, whereas chroma and saturation are relative ones. Distinguishing saturation from chroma is subtle. Decreasing chroma fades a color to gray of equal lightness, as shown in the slices on the Munsell Color Tree in Figure 3. Saturation is a property of objects, and remains constant as the intensity of the light changes, even if the color changes in chroma.

To understand these terms better, consider the following example: Take a brightly colored object such as a bright red ball. It is a saturated red hue all over, independent of the changes in color caused by the play of light and shadow upon it. That is, hue and saturation are inherently defined by the object color, not its lighting. Across

36

the shaded surface of the ball, the lightness and chroma will vary. Make the lighting twice as bright, and the ball will appear more colorful and brighter. Dim the lighting and it will become less so. But, the lightness and chroma will remain the same because they are relative to the light.

The goal of a color appearance model is to take some sort of measurement, such as CIE tristimulus values for the color, its surrounding colors and the light source, and produce perceptual attributes such as hue, colorfulness, and brightness. This process is summarized in Figure 15. These attribute values can then be used to match colors across different viewing conditions. As a simple example, given a color on one background, compute the color that would look the same on a different background. A more complex example is to transform an image on a monitor such that it "looks the same" on a print. This is much more difficult because images evoke our knowledge of the world, creating situations where gray can appear black or white, depending on context as shown in Figure 1. Plus, the colors needed to "look the same" may not be available on the target medium, so the problem becomes not one of matching, but of creating the best approximation, as will be discussed further in the chapters on color reproduction (Chapters 5–9).

A color appearance model must consider all the colors in an image together, rather than as individual colored pixels. This is essential to model spatially defined appearance effects such as simultaneous contrast. The minimum amount of spatial information included in color appearance models is the color itself (called the *stimulus*), its immediate surrounding color (the *background*), and the larger viewing environment (called the *surround*), as summarized in Figure 16. Typically, the stimulus and the background are specified as tristimulus values, whereas the surround is one of dark, dim, or normal.

All color appearance models include a way to compensate for both brightness adaptation and chromatic adaptation. Even CIELAB implements a simple adaptation model, where the tristimulus values of each color are divided by the tristimulus values of the white reference. This maps white to white, but does not accurately transform other colors, causing unexpected hue shifts.

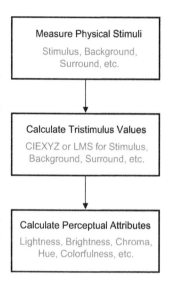

Figure 15.

The process of color appearance modeling. Measurements such as CIE XYZ are transformed to appearance phenomena such as hue, brightness, colorfulness, etc.

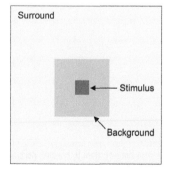

Figure 16.

Spatial terms used in color appearance models.

Figure 17.
Formulation for a von Kries chromatic adaptation transformation. The matrix values are those recommended for CIECAM02.

$$\begin{bmatrix} X_2 \\ Y_2 \\ Z_2 \end{bmatrix} = A^{-1} \begin{bmatrix} L_{W2}/L_{W1} & 0.0 & 0.0 \\ 0.0 & M_{W2}/M_{W1} & 0.0 \\ 0.0 & 0.0 & S_{W2}/S_{W1} \end{bmatrix} A \begin{bmatrix} X_1 \\ Y_1 \\ Z_1 \end{bmatrix}$$

$$A = \begin{bmatrix} 0.7328 & 0.4296 & -0.1624 \\ -0.7036 & 1.6975 & 0.0061 \\ 0.0030 & 0.0136 & 0.9834 \end{bmatrix}$$

A better, and almost equally simple transformation is a von Kries transformation. A von Kries transformation is computed from the ratio of the old and new white, expressed as cone response functions. For example, if the white shifts such that the blue component changes by some factor k, the blue cone response will change by 1/k. The more blue the light, the less sensitive the blue cone, and vice versa.

The process is summarized in Figure 17, where X, Y, and Z are tristimulus values; L, M, S are cone responses; and A is a matrix that transforms between them. Given a set of tristimulus values that define the old ($L_{w1}$, $M_{w1}$, $S_{w1}$) and new ($L_{w2}$, $M_{w2}$, $S_{w2}$) values of white, first use a matrix multiplication to convert from tristimulus values to cone response values. Then, compute the ratios in LMS space and convert back to tristimulus space using the inverse of the XYZ to LMS matrix. The ideal matrix for this transformation is a current topic of study, but the one shown in the figure is the one proposed for the most recent CIE-sponsored color appearance model, CIECAM02.

One aspect of color appearance modeling that has an important impact on image reproduction is the effect of the surround, which is the overall brightness of the viewing environment. The effect of changing the surround in image reproduction is to change the perceived contrast. Making the surround darker decreases the perceived contrast; to compensate, the image contrast must be increased. For example, given an image captured under some level of illumination, such as an outdoor scene, it is necessary to increase the contrast of the reproduced image if the viewing environment is less bright, such as a movie theater.

Reference.
Nathan Moroney, Mark D. Fairchild, Robert W.G. Hunt, Changjun Li, M. Ronnier Luo, and Todd Newman. "The CIECAM02 Color Appearance Model." IS&T/SID 10th Color Imaging Conference (2002).

This result was demonstrated by Bartleson and Breneman in the late 1960s, and is regularly applied in image reproduction today. Their results are summarized in Figure 18, which is a log-log plot of the original scene luminance values versus the corresponding reproduced luminance values. The slope of the plot corresponds to the power, or *gamma*, of the corresponding function (remem-

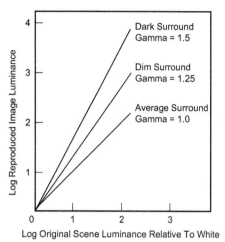

Figure 18.
Bartelson and Breneman recommendations for increasing contrast as a function of surround.

ber that a power function $x^\gamma$ plotted on a log-log scale will be a straight line with slope = $\gamma$). If the viewing environment is "average," which means it matches that of the original scene, no compensation is needed and gamma = 1.0. A dim surround is typical for television and monitor viewing, a dark surround for movies and slides. Each requires increasing the contrast, as shown in the figure, to achieve the same overall appearance as the original. These results are part of the practice in professional image reproduction industries such as television and graphic arts, and are encoded in some form in most color appearance models.

# Summary

The appearance of color is much more complex than the simple, additive models of trichromatic matching. The dimensions of perceptual color spaces are not red, green, and blue, but lightness, hue, and chroma. This reflects the opponent color encoding performed early in the processing of color signals coming from the eye. Another important factor in color appearance is adaptation, which means that the visual response changes, adjusting for a wide range of brightness levels and reference illumination.

Reference.
C.J. Bartleson and E.J. Breneman. "Brightness Perception in Complex Fields." *Journal of the Optical Society of America* 57 (1967), 953–957.

References.

Zhang and Wandell. "A Spatial Extension of CIELAB for Digital Color Image Reproduction." *SID Journal* (1997).

M.D. Fairchild and G.M. Johnson. "The iCAM Framework for Image Appearance, Image Differences, and Image Quality." *Journal of Electronic Imaging,* submitted (2003).

The appearance of a simple color patch is strongly affected by its size and its background. Combine colors in a scene, and the interactions become even more complex. The visual system is designed to view objects in a world of ever-changing light and shadow. When colors are associated with objects, their appearance becomes influenced by our experience in the world, adding cognitive effects to the perceptual effects of color appearance.

Color appearance models try to capture this complexity and apply it to applications such as digital imaging. The simplest color appearance models are the color difference spaces, such as CIELAB. These organize color perceptually, modeling adaptation (but not very accurately) by including a reference white. They do not, however, include the effect of surrounding colors. The CIE codified a more complete model in 1997, called CIECAM97s, which is still under active development. The most recent improvements have been proposed as CIECAM02. Other models include RLAB and LLAB (improved CIELAB), Hunt, Nayatani, Guth, and ATG. All but CIECAM02 are discussed in Fairchild's book.

Another body of work in this area more directly addresses image appearance, and the automatic generation of important spatial information from imagery. The S-CIELAB specification created at Stanford by Brian Wandell's group, and the iCAM work by Fairchild and his students at RIT are examples of this form of image appearance (in contrast to color appearance) model. Given how fast this domain is changing, websites and conference papers are the best references. Many of the IS&T sponsored conferences, but especially the Color Imaging Conference and the Electronic Imaging Conferences, are the places where work in the area of color appearance and its models are published.

# 2 Color Appearance

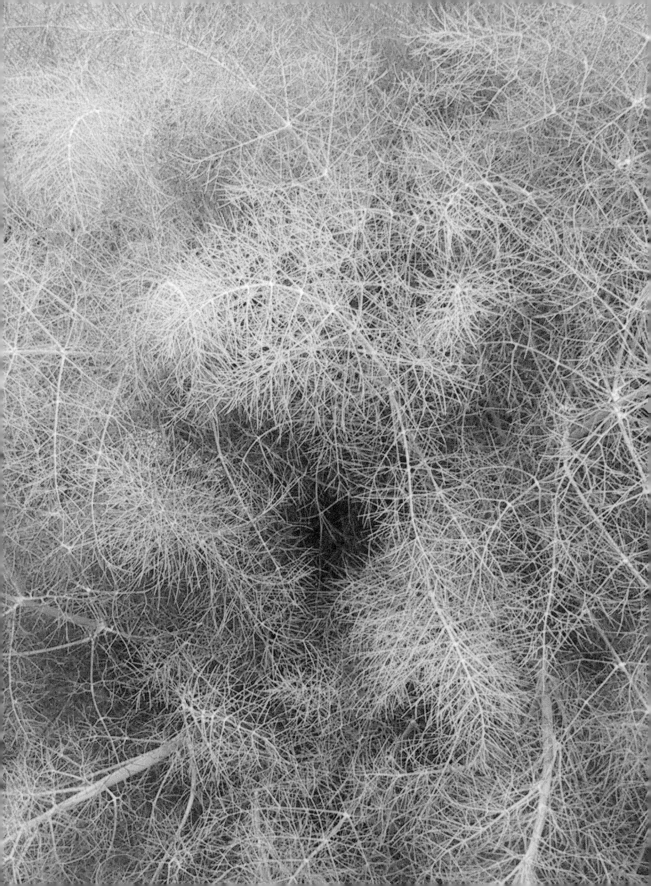

# 3

# RGB and Brightness— Fundamental Concepts

Color described as numerical combinations of red, green, and blue intensity values is fundamental to digital color. It is the encoding for colored pixels in images, and the mechanism for specifying the colors produced on a color display. Defining a brightness scale as numerical steps from dark to light is similarly fundamental to digital color. This chapter describes, in detail, RGB color and numerically defined brightness scales, emphasizing the use of colorimetric measurement to provide a quantitative, colorimetric foundation for such specifications. These concepts will appear again and again throughout the rest of this book.

# Introduction

Digital color images are encoded as triples of red, green, and blue pixel values. Key to specifying and controlling RGB color is defining what these numeric values mean in physical and in perceptual terms. A truly perceptual specification is not simple, as discussed in the previous chapter. However, pixel values can be defined in colorimetric terms, such as CIE tristimulus values (CIE XYZ), which can be measured directly and accurately with color measurement instruments.

Digital color displays create color by adding red, green, and blue light through either phosphor emissions in CRT (monitor) displays or by light shining through colored filters in flat panel displays or digital projectors. Varying the brightness of the red, green, and blue primary colors varies the color produced. The set of all colors produced on a display is called the *color gamut* for that display. Different displays have different gamuts, depending on their primary colors. These gamuts can also be defined with respect to measurable quantities, such as CIE tristimulus values (CIE XYZ).

Take a particular color and change only its brightness to create a sequence from dark to light, such as the sequence that steps from black, through shades of gray, to white. In this chapter, the definition of such sequence is called a *brightness scale*. In other contexts, such a scale might be called a *lightness scale*, or an *intensity transfer function,* or even a *gamma function*, as one common form of brightness scale is based on power functions of the form $y = x^\gamma$.

The chapter starts with a description of RGB color spaces and their properties. It then describes how such a color space can be defined colorimetrically, by creating a mapping from RGB to CIE XYZ. These concepts are then used to model a color display system in terms of colorimetric properties. The following section provides more detail about brightness scales, including the various nonlinear functions of brightness most commonly used in digital imaging and how they are applied. The chapter concludes with a detailed description of the sRGB color space, as this combines all the basic principles into one practical example, complete with formulas.

# RGB Color Spaces

An RGB color space consists of three primary colors, red, green, and blue, that are combined according to the principles of additive mixture. In terms of physical lights, adding colors means that their spectra combine; the resulting color is what a person with normal vision would see.

In the digital color world, we rarely think of the spectra associated with the RGB color primaries. A color is a triple of numbers, one associate with each primary, that combine by adding them as if they were vectors. That is, add red to red, green to green, and blue to blue. For example $(0, 1, 0.5) + (1, 0, 0.2) = (1, 1, 0.7)$. In mathematical terms, the R, G, and B form the axes of a three-dimensional color space.

An RGB color space can be described as a unit color cube, with black at $(0, 0, 0)$ and white at $(1, 1, 1)$. The other corners of the cube are the *primary colors* (red, green, and blue) and the pairwise sums of the primaries, sometimes called the *secondary colors* (cyan, magenta, and yellow). These colors and their color values are shown in Figure 1. Mathematically, colors can fall outside of the unit cube simply by including values greater than 1 or less than 0. Most applications that use RGB color spaces restrict the values to the range $(0, 1)$. The unit cube, therefore, is the gamut of colors defined for that color space.

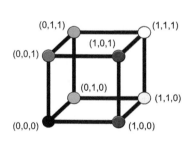

Figure 1.

The RGB color cube, with each vertex labeled with its corresponding RGB triple.

Equal values of red, green, and blue produce gray, and lie along the main diagonal of the cube (running from black to white). For any RGB triple, the smallest value defines the amount of gray added to the color, and the other two primaries define the hue and degree of saturation. For example, the color $(0.7, 0.3, 1.0)$ is a purple equivalent to $(0.4, 0.0, 0.7)$ with $(0.3, 0.3, 0.3)$ added to it. Fully saturated colors contain no gray and lie on the surface of the cube.

These properties can be used to create an interpretation of the RGB color cube organized as a perceptual color space. The main diagonal of the cube becomes the achromatic axis running from black to white. The distance from this axis becomes the "saturation," and hue is arranged in some form of closed shape, most often a hexagon, triangle or circle. Such spaces will be described in more detail in Chapter 11, which discusses the HSV and HLS models for color selection tools. Such spaces are not true perceptual spaces, but simply renotations of the RGB color cube.

# RGB and Colorimetry

The color cube is an abstraction. To create visible colors, we must bind the primaries to physical colors, such as colored lights or the colored phosphors of a CRT display. Varying the intensity of these lights creates a full gamut of colors, filling the volume of the cube. For a physically defined RGB color space, there is always a finite color gamut defined by the limits of the primary colors.

The light from these physically specified primaries can be quantitatively defined using the color measurement techniques based on CIE colorimetry, as described in Chapter 1. This quantitative description can be used to predict any combination of the primaries using Grassmann's additivity law, which was also described in Chapter 1. This law states that given the tristimulus values for the primary colors, we can compute the tristimulus values for any mixture by summing the tristimulus values of the primaries. For example, let the tristimulus values for red be $X_R$, $Y_R$, and $Z_R$, and similarly for green and blue. For some color, C, whose RGB specification is (0.5, 1, 0.2):

$$
\begin{aligned}
X_C &= 0.5X_R + 1.0X_G + 0.2X_B \\
Y_C &= 0.5Y_R + 1.0Y_G + 0.2Y_B \\
Z_C &= 0.5Z_R + 1.0Z_G + 0.2Z_B.
\end{aligned}
$$

This result describes a linear transformation, which can be described as $3 \times 3$ matrix, that converts from RGB to XYZ, and whose inverse converts from XYZ to RGB. This transformation can be used both for specification, and for conversion between RGB color spaces. More detailed mathematics for this transformation are provided at the end of the chapter. For those not interested in the mathematical details, simply note that there exists a $3 \times 3$ matrix, defined by the tristimulus values of the primaries, that can be used to convert any RGB color into its associated tristimulus values.

The CIE tristimulus values also define a three-dimensional space, whose axes are X, Y, and Z. The result of applying a $3 \times 3$ matrix as described above to the corners of a physically defined RGB color cube is shown in Figure 2. The primary colors become vectors of different lengths, proportional to their brightness. Their direction is defined by their color, specified as an XYZ triple. White is a color specific to these primaries, defined by summing them. Black maps to (0, 0, 0).

The color cube retains a roughly cubical shape with three pairs of parallel sides, but its square faces are now parallelograms. There is still an achromatic axis along the main diagonal, and the fully saturated colors are still on the surface, which encloses the color gamut. For those familiar with 3D transformation matrices, this shape is a visual confirmation of the affine transformation (a combination of rotation, scaling, and skewing) described by the $3 \times 3$ matrix.

Plotting an RGB color space on the chromaticity diagram by converting the CIE XYZ coordinates to chromaticity values produces a triangle, which is also shown in Figure 2. The corners are the primary colors. The secondaries lie along the line between the primaries at a distance proportional to the relative brightness of the two primaries. White is marked with a solid dot, and generally falls near the center of the triangle. Black is directly under white—this is effectively the skewed cube in CIE XYZ coordinates projected along its black/white axis. The lines running from white to the colors are the projections of the edges of the color cube—half are connected to black, half to white. All colors in the color gamut fall inside the triangle.

RGB color spaces defined by different primaries project as different triangles. Comparing these projections can be useful, but does

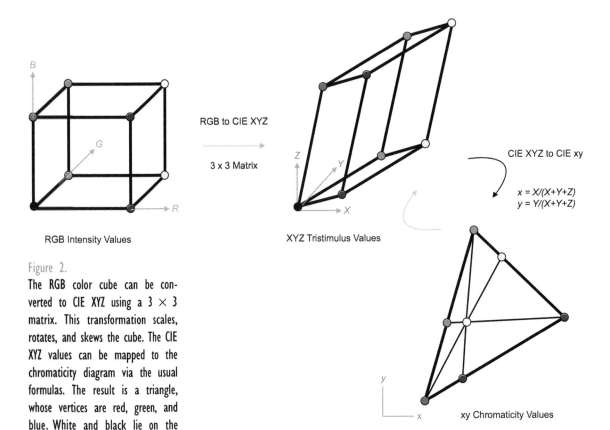

RGB to CIE XYZ

3 x 3 Matrix

RGB Intensity Values

XYZ Tristimulus Values

CIE XYZ to CIE xy

$x = X/(X+Y+Z)$
$y = Y/(X+Y+Z)$

xy Chromaticity Values

Figure 2.
The RGB color cube can be converted to CIE XYZ using a 3 × 3
matrix. This transformation scales,
rotates, and skews the cube. The CIE
XYZ values can be mapped to the
chromaticity diagram via the usual
formulas. The result is a triangle,
whose vertices are red, green, and
blue. White and black lie on the
same point.

not provide complete information about whether a color lies inside
both color spaces because of the missing brightness information.
That is, a color that falls inside the triangle may not actually be
inside the color space volume because it is too bright or too dark. An
accurate comparison can only be made in a three-dimensional space.

## Numeric Brightness Scales

Take the numbers from 0 to 255 and create a brightness scale that
has some concrete meaning. This seemingly simple problem has generated more confusion and outright "holy wars" than any other in

digital color. Suitably refined experts can even manage to disagree on the results of measuring physical systems, especially when trying to give these results perceptual significance.

Brightness can be measured as either intensity or luminance. As described in Chapter 2, intensity is a measure of the total light energy in the spectrum, and luminance is a measure of this energy weighted by how the eye responds to different wavelengths. For any given spectral distribution, therefore, luminance is a fixed fraction of intensity. If changing the brightness of a color only involves scaling its spectrum, its luminance scale will always be a scaled version of its intensity scale. Specifications for brightness scales are usually normalized to some standard maximum value so that all these scale factors drop out, making relative luminance the same as relative intensity in many contexts.

When comparing brightness values for different hues, however, there is a difference between luminance and intensity, where the luminance value is more indicative of the perceived brightness. In general, when measuring brightness with respect to color, it is better to specify luminance values.

The simplest brightness scale to specify is one that is linear in relative luminance (or intensity). However, such a scale does not make efficient use of the limited number of encoding steps that can be specified with integer pixels. In digital color, pixel values are most often stored as 8-bit values, which represent the numbers between 0 and 255.

For 8-bit pixels on a typical display, there will be quantization steps or contour lines visible in the dark colors if the encoding is linear with respect to luminance. The solution is to define a nonlinear encoding that is a better map to perception. This means defining the brightness scale so that each step is just noticeably different from the next, resulting in a perceptually uniform scale.

Unfortunately, there is no simple definition of a perceptually uniform brightness or lightness scale. The perception of uniformity depends on many factors, including the surrounding colors, as discussed in Chapter 2. Figure 3 emphasizes this point—the three scales are identical, but the perceived spacing of the gray patches varies with

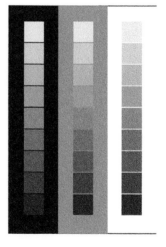

Figure 3.

Demonstration that the perception of uniform brightness steps changes with the surrounding colors. The three grayscales are identical, and are designed to be equal steps in reflectance.

background. The scale in this example should approximate uniform steps in reflectance (reflected light), which are equivalent to uniform steps in intensity. On each scale, the patch that seems to be about halfway between black and white changes with the background. Note also that there are more dark colors, relative to this middle gray than light ones.

Because there is no single definition of "perceptually uniform," there is no standard perceptual encoding for brightness used in digital color systems. In electro-mechanical systems such as cameras and displays, there can also be engineering constraints that influence the implementation of the encoding and decoding, especially when applied to video. Figure 4 shows several common non-linear functions for encoding relative luminance as pixels. While similar in shape, they can produce visibly different results to the trained eye when applied to appropriate images. The flip side of this statement is that to the casual user, they are equivalent as long as some non-linearity of this form is used.

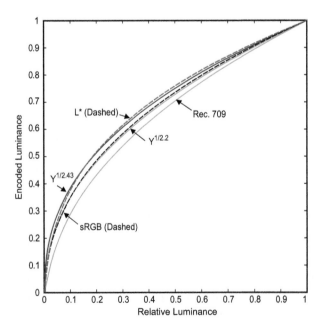

Figure 4.

Several non-linear brightness encodings, including: sRGB, the ITU Rec 709 video encoding, the perceptual brightness scale L*, and two power functions.

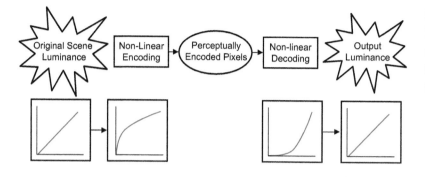

Figure 5.
The process of using a non-linear encoding for image reproduction. The small graphs illustrate the brightness mapping at each step.

Figure 5 summarizes the process of using a non-linear encoding for luminance in a digital color reproduction system. The small graphs beneath each step indicate the brightness mapping at each step. The original scene luminance values, which are linear, are non-linearly encoded to create pixels, whose numeric values specify uniform perceptual steps in luminance. For example, imagine a camera pointed at a scene, which maps the normalized scene intensity values it captures into pixel values using an encoding function like one of those shown in Figure 4. Decoding reverses the process by mapping the pixels through the inverse of the encoding function, for example, to reproduce the original (linear) relative luminance values on a display.

Figure 6 shows in detail a pair of encoding and decoding functions based on a simple power function and its inverse. The dashed lines highlight that the value, perceptually half-way between black and white (encoded as a 0.5), represents a relative intensity of approximately 0.22. This indicates that more pixels must be assigned to darker colors than to light ones in a perceptually uniform encoding, as was suggested by the discussion of Figure 3.

Those familiar with display technology may recognize the decoding function in Figure 6(b) as the CRT "gamma function," which maps voltage to intensity (or luminance). This function takes the form $Y = V^\gamma$, where $Y$ is the displayed luminance, $V$ is the input voltage (equivalent to pixel value), and $\gamma = 2.2$. This means the CRT hardware will automatically decode image pixels that have been encoded by raising them to the 1/2.2 power, a property that is ex-

Figure 6.

A typical encoding function (a) and decoding function (b), based on a simple power function.

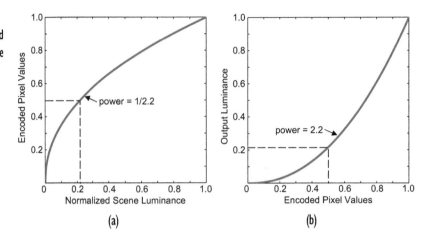

(a)                    (b)

ploited by all video encodings. While only CRTs inherently have this non-linear relationship between voltage and luminance, most other display systems mimic it for compatibility with video.

For graphics and imaging applications, pixels are created algorithmically, rather than captured. If these pixels are displayed directly through the CRT's non-linear function, without any form of "gamma correction" (see Chapter 10), their values are a non-linear, perceptually uniform encoding of luminance, just like a video image.

## Linear and Non-Linear RGB Color Spaces

The transformation from RGB to CIE XYZ described earlier and illustrated in Figure 2 tacitly assumed that there was a linear mapping from RGB color values to luminance. That is, (0.5, 0.5, 0.5) describes a color whose measured luminance is 1/2 that of (1, 1, 1). Such a space is called a *linear* RGB color space.

If, however, the color values are non-linearly encoded (as described in the previous section) the color space is a *non-linear* RGB color space. To map a non-linear RGB color space into CIE XYZ requires an additional step before applying the 3 × 3 matrix. The pixel values must be converted to linear intensity values using a decoding function as described in the previous section. The inverse transform from CIE XYZ to non-linear RGB must encode the color values after applying the matrix—this is summarized in Figure 7.

In a linear RGB color space, all colors specified as equal values of R, G, and B (often called the *gray axis* for an RGB color space) lie

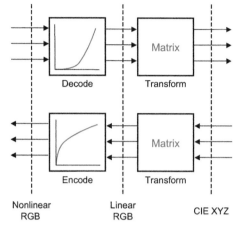

Nonlinear RGB     Linear RGB     CIE XYZ

Figure 7.

The process of transforming a non-linear RGB color space to CIE XYZ, and vice versa, which involves an encoding and decoding function as well as the matrix multiplication.

along the vector from black to white. All the values in the gray axis contain RGB in the same relative proportions, and all map to the same chromaticity coordinate. For a non-linear RGB color space, this will only be true if the encoding functions for the primaries are identical, and hence identical to the one for the gray axis. If the functions vary, the single curve shown in the encoding/decoding steps in Figure 7 must be three separate curves to produce an accurate transformation. As a result, the values of the gray axis will not lie along the same vector, and will not plot to the same point on the chromaticity diagram.

## RGB Color Spaces and Displays

CRT displays provide a physical model of a non-linear RGB color space, and one that is familiar to most readers of this book. The primary colors are defined by the phosphors. There is a non-linear mapping from pixel values to luminance, defined by the display electronics. We assume for this discussion that the primaries act as entirely independent light sources, and do not vary in color as they change in intensity. This is essentially true for CRTs, but not necessarily for other display technologies, as will be discussed further in Chapter 7.

The process of defining the transformation between RGB pixels and CIE XYZ for a display system is called *characterizing* the display. The broader topic of characterizing displays and other digital color output systems will be discussed in the chapters on color reproduction and management (Chapters 5–9). The primary goal here is to provide more insight into RGB color spaces by discussing them in the context of a familiar physical system.

A characterization consists of a $3 \times 3$ matrix, plus a set of tables that map pixel values along each primary to luminance values. These are created by taking measurements of the display's primary colors. The characterization can be inverted (to produce the CIE XYZ to RGB mapping) by inverting the matrix and the tables.

The values needed for the $3 \times 3$ matrix are the CIE XYZ values for each color primary displayed at its maximum brightness. To get these values, one must measure each primary color at its full intensity with a colorimeter or spectroradiometer. The mapping from pixel values to luminance (Y) is defined by measuring in steps along each primary. Note that the chromaticity values (x, y) should be constant for each primary as the brightness changes.

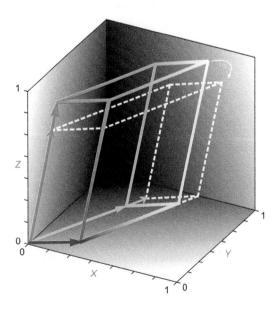

Figure 8.

The RGB color space for a display with two different settings for the brightness of green and blue (solid versus dashed lines). Note how the value for white changes.

The chromaticity of the primary colors is defined by the CRT phosphors and does not change over time. The maximum luminance for each primary can vary, changing the size and shape of the resulting color gamut, including the definition of white. This is illustrated in Figure 8, which shows an RGB color space for a display at two different RGB brightness settings, plotted in CIE XYZ: the solid lines are the original settings; the dashed lines show the result of increasing the brightness of green and decreasing the brightness of blue. Their sum, as expected, changes, creating a different gamut and a different color for white. Figure 9 shows Figure 8 projected onto a chromaticity diagram. The vertex colors for the original settings are shown as filled circles, and for the new settings as squares. The corners of the triangle (the chromaticity coordinates of the primaries) do not change, but the white point and the positions of the secondary colors have shifted. For example, cyan has moved towards the stronger green primary.

As was discussed above, a non-linear RGB color space is not guaranteed to map its grayscale along a single vector, and hence to a constant chromaticity. Measuring each primary separately, rather

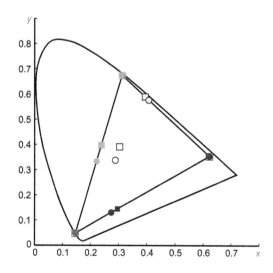

Figure 9.
Chromaticity view of Figure 8. The position of the primary colors has not changed, but that of white and the secondary colors has shifted.

than relying on a single pixel-to-luminance function defined by measuring the gray axis, will create a valid RGB to CIE XYZ transformation in all cases. Perceptually, a gray that smoothly shifts color a bit is not very noticeable, especially as hue is difficult to perceive in dark colors.

In the idealized RGB color spaces discussed in this chapter, black is mapped to XYZ = (0, 0, 0). This is physically equivalent to a black hole—no light shines out. In real systems, black is a physical color and is rarely exactly the same color as white. The color of black acts as an offset to the 3 × 3 matrix transformation, but is only included in the characterization if the black is significantly bright, as in some digital projectors. Including the values for black in the transformation converts the 3 × 3 matrix to a 4 × 4, the details of which are discussed at the end of the Chapter. On a CRT display, the black is so dark that it is often difficult to measure accurately, so black is conveniently assumed to be (0, 0, 0).

# A Standard RGB Color Space (sRGB)

The sRGB specification is a standard specification for RGB color images defined in terms of a virtual display. It is a non-linear color space that has been standardized by the International Electrotechnical Commission (IEC) as IEC 61966-2-1, and has been gaining acceptance as the default RGB color space for many applications, including: web imaging; image editing programs such as Adobe Photoshop; as the default color space for Microsoft Windows color management; and as the default input specification for many desktop printers.

The sRGB specification includes the phosphor chromaticities, the white point and the maximum luminance of a "standard display." There is a single, non-linear mapping that is applied to all three primaries. This is sufficient to

|   | Red | Green | Blue | White (D65) |
|---|---|---|---|---|
| x | 0.6400 | 0.3000 | 0.1500 | 0.3127 |
| y | 0.3300 | 0.6000 | 0.0600 | 0.3290 |

Table 1.
The sRGB color specification.

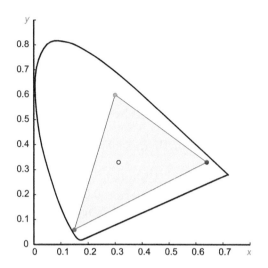

fully characterize the RGB color space with respect to CIE tristimulus values. The sRGB specification also specifies that the images will be viewed in a normal office environment to try to include some aspects of image appearance. This affects the interpretation of the transfer curve encoding, and has caused the most difficulty and confusion about sRGB.

The chromaticity coordinates for the sRGB space are shown in Table 1 and plotted on the chromaticity diagram in Figure 10.

The maximum luminance (Y) for sRGB white is 80 candellas/meter$^2$. This, plus the chromaticity coordinates, is sufficient to compute the matrix values that transform between CIE XYZ and RGB. The details of this computation are given in the next section.

Like many non-linear encoding functions, sRGB specifies a brightness encoding that is a combination of a power function and a linear segment near zero. Linear RGB values are converted to non-linear RGB$_{sRGB}$ values by:

$$\text{If R, G, B} > 0.0031308$$

$$R_{sRGB} = 1.055R^{(1.0/2.4)} - 0.055$$
$$G_{sRGB} = 1.055G^{(1.0/2.4)} - 0.055$$
$$B_{sRGB} = 1.055B^{(1.0/2.4)} - 0.055$$

$$
\begin{aligned}
\text{else} \\
R_{sRGB} &= 12.92R \\
G_{sRGB} &= 12.92G \\
B_{sRGB} &= 12.92B
\end{aligned}
$$

The non-linear values are decoded by:

$$
\text{If } R_{sRGB}, G_{sRGB}, B_{sRGB} > 0.04045
$$

$$
\begin{aligned}
R &= ((R_{sRGB} + 0.055)/1.055)^{2.4} \\
G &= ((G_{sRGB} + 0.055)/1.055)^{2.4} \\
B &= ((B_{sRGB} + 0.055)/1.055)^{2.4}
\end{aligned}
$$

$$
\text{else}
$$

$$
\begin{aligned}
R &= R_{sRGB}/12.92 \\
G &= G_{sRGB}/12.92 \\
B &= B_{sRGB}/12.92
\end{aligned}
$$

The 2.4 power in the sRGB specification is misleading, because the linear segment requires adding a scale and offset to the power function. As a result, the simple power curve that most nearly approximates the sRGB encoding has an exponent of 1/2.2 (2.2 for decoding). This is shown in Figure 11, which is a graph of values in the range (0, 1) transformed through the sRGB encoding function compared to a graph of the same values raised to the 1/2.2 power. The difference between the two curves is extremely small.

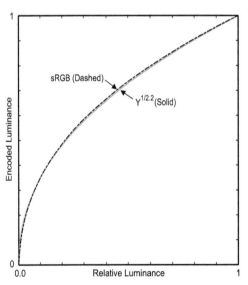

Figure 11.
The sRGB encoding function (dashed), compared to a simple power function whose exponent is 1/2.2 (solid).

An sRGB-compatible display, therefore, can be approximated by one whose phosphors and white point match the specification, and whose "gamma function" is $Y = V^{2.2}$. This approximation will be used throughout this book when discussing sRGB.

# Math for RGB Color Spaces

The $3 \times 3$ matrix that transforms between linear RGB and CIE XYZ is defined directly from the tristimulus values of the primaries. If $X_R$, $Y_R$, $Z_R$ are the tristimulus values of R at its maximum value, and similarly for G and B; and black maps to $XYZ = (0, 0, 0)$ then the transformation is

$$\begin{bmatrix} R & G & B \end{bmatrix} M = \begin{bmatrix} X & Y & Z \end{bmatrix},$$

where

$$M = \begin{bmatrix} X_R & Y_R & Z_R \\ X_G & Y_G & Z_G \\ X_B & Y_B & Z_B \end{bmatrix}.$$

As in the sRGB specification, rather than specifying the tristimulus values for each primary, it is more common to specify the chromaticity coordinates of the primary colors and the white point plus the maximum luminance of white ($Y_W$). To construct the tristimulus values from this specification, follow these steps:

1. Compute $z$ for each primary and white using the property of chromaticity values that $x + y + z = 1$.

2. Compute the tristimulus values ($XYZ_W$) for white from $x_W, y_W, z_W, Y_W$ as follows:

   a. $S_W = Y_W / y_W$, where $S_W = (X_W + Y_W + Z_W)$

   b. $X_W = x_W S_W$, $Y_W = y_W S_W$, $Z_W = z_W S_W$.

3. Use the property that $XYZ_W = XYZ_R + XYZ_G + XYZ_B$ to set up the following set of simultaneous equations:

$$x_R S_R + x_G S_G + x_B S_B = X_W$$
$$y_R S_R + y_G S_G + y_B S_B = Y_W$$
$$z_R S_R + z_G S_G + z_B S_B = Z_W.$$

4. Solve for $S_R, S_G, S_B$ using any appropriate method.

5. Compute the tristimulus values for the primaries from $X_R = x_R SR, \ Y_R = y_R S - R$, etc.

The resulting values for an sRGB monitor are given in the Table 2.

Table 2.

Normalized tristimulus values for an ideal sRGB display.

|  | X | Y | Z |
|---|---|---|---|
| red | 0.4124 | 0.2126 | 0.0193 |
| green | 0.3576 | 0.7152 | 0.1192 |
| blue | 0.1805 | 0.0722 | 0.9505 |

All of these values have been normalized so that the luminance of white is 1.0 ($Y_w = Y_R + Y_G + Y_B$). Simply scale all the XYZ values to normalize differently. For example, to match the specification for sRGB, scale by 80 candellas/meter$^2$.

The inverse of the matrix M transforms from CIE XYZ to linear RGB.

$$[ \ X \quad Y \quad Z \ ]M^{-1} = [ \ R \quad G \quad B \ ]$$

To convert from one RGB color space to another, you can concatenate these two transformations. If $M_1$ is the matrix that transforms $RGB_1$ to CIE XYZ, and $M_2$ is the matrix for $RGB_2$, then:

$$RGB_2 = RGB_1 M_1 M_2^{-1},$$

If black is not set to XYZ = (0, 0, 0), then it must be included as an offset in the transformation. This can be most simply achieved by using homogeneous coordinates, making the matrix a 4 × 4 as shown: where $X_K$, $Y_K$, $Z_K$ are the tristimulus values for black. This matrix

$$[\ R \quad G \quad B \quad 1\ ] \begin{bmatrix} X_R - X_K & Y_R - Y_K & Z_R - Z_K & 0 \\ X_G - X_K & Y_G - Y_K & Z_G - Z_K & 0 \\ X_B - X_K & Y_B - Y_K & Z_B - Z_K & 0 \\ X_K & Y_K & Z_K & 1 \end{bmatrix} = [\ X \quad Y \quad Z \quad 1\ ],$$

can then be used as above. This convenient formulation for managing non-zero black was introduced in a paper on characterizing projection displays.

For non-linear RGB spaces, all RGB values must be decoded to create linearly spaced intensity values before the RGB to CIE XYZ matrix (either 3 × 3 or 4 × 4) is applied. Similarly, the RGB results produced by the CIE XYZ to RGB matrix transformation are linear intensity values that must be encoded to create a non-linear representation. In practice, the encoding and decoding functions are usually implemented as tables.

# Summary

Representing color by three numeric components (red, green, and blue) is fundamental to digital color. These numbers take on a physical meaning when they are used to drive a display. To create a definition independent of specific display hardware, they can be measured and characterized based on colorimetric principles to define a unique transformation between the RGB pixel values and CIE tristimulus values (CIE XYZ).

Brightness scales are commonly used to define the sequence of colors that lie between black and white, or more generally, between black and some maximally bright color. Like RGB values, they need a metric

Reference.
M. Stone. "Color Balancing Experimental Projection Displays." In *Proceedings of the 9th Color Imaging Conference: Color Science, Systems, and Applications* (2001), pp. 342–347.

specification, such as intensity or luminance, to define them precisely.

An RGB color space is a volume of colors specified as triples of RGB values. Converting an RGB color space into the three-dimensional space defined by CIE XYZ transforms this cube by mapping it through a 3 × 3 matrix transformation. The result is a cube that has been rotated, scaled, and skewed.

The RGB color values in a linear RGB space represent steps in intensity. A non-linear RGB space encodes the values through a non-linear function, usually to make the encoding more perceptually uniform. Most image encodings in digital color are based on non-linear RGB color spaces. Non-linear color values must be converted to linear ones before transforming them through any matrix.

The mathematics at the end of this chapter are basic linear algebra, which should be familiar to many in the graphics and engineering communities. I include it here for the convenience of implementers of graphics systems, and for those for whom the equations speak louder than words. Qualitatively, all that is important to remember is: 1) there is always a matrix defined by the specific RGB involved; and 2) there is always a brightness mapping, which must be precisely defined for a precise color specification. In digital color systems, variation in the brightness mapping causes more problems with color fidelity than variation in the RGB specification, especially across display systems.

# 3 RGB and Brightness—Fundamental Concepts

# 4
# Color in Nature

What physical mechanisms produce the colors around us? Models of physical color are simulated in the field of computer graphics, and are essential to the production of the inks and dyes in print and film—plus, they are fascinating. The previous chapter described digital color, represented by numbers and additive mixture. This chapter describes various physical mechanisms for producing color, including absorption, refraction, and different types of scattering. It also introduces paints and dyes, the technologies for manufacturing color in the physical world.

## Introduction

The process of seeing color starts with a spectral distribution entering the eye. But where do these spectra come from? Some are directly viewed light, such as the color of a laser pointer, a glowing light bulb, a display of fireworks, or a neon sign (Figure 1). However, most are the result of light shining on or through objects, whose

Figure 1.

The glowing gasses of a neon sign emit light to produce color.

65

Figure 2.

Light shining on or through an object reveals its color.

Figure 3.

Interference creates the colors on a soap bubble's surface. (Copyright Sterling Johnson, the Bubblesmith.)

surfaces selectively absorb, reflect, or transmit different wavelengths (Figure 2). The resulting spectra define their colors. Other colors, including some of the most vivid, are caused by interference and scattering, effects that selectively enhance or destroy specific spectral colors.

The swirling rainbows on a soap bubble or oil slick are called *iridescence*, and are caused by light reflecting from two surfaces of a thin film such that the reflections interfere with each other, like wave patterns on water. The result is that some wavelengths are amplified and others are minimized. The patterns swirl and change color as the thickness of the thin film varies (Figure 3). Interference also accounts for the vivid, jewel-like colors of certain feathers, butterfly wings, and beetles, where fine structures selectively enhance and destroy different colors. It also creates the opalescent colors of mother-of-pearl and opals themselves.

*Scattering* creates the vivid colors of tropical fish, the blue-green of glacier lakes, brilliant white clouds, and the deep blue sky (Figure 4). Small particles scatter some wavelengths back toward the viewer, and the rest are absorbed by the material behind them. It is the combination of wavelength-specific scattering and absorption of the remaining colors that gives such vivid results. Scattering also occurs with larger particles. The color created depends on the reflectance and absorption properties of the material in the particles. Particles that reflect all the light appear white.

The colors in the sky are created by the light of the glowing sun interacting with the atmosphere—light striking molecules and small particles is scattered. The more a wavelength is scattered, the more likely it is visible. For example, the sky is blue because the atmosphere scatters more blue light than other colors. The vivid colors of the rainbow (Figure 5), however, are created by water droplets acting as prisms. They refract (bend) the light as a function of its wavelength, spreading its colors in a spectacular display, an effect called *dispersion*.

Color can be manufactured and applied. Many colors are created by fine particles known as *pigments*, which, when suspended in a medium, selectively scatter and absorb different colors. Paint is a suspension of pigments that can be applied to surfaces. A colorant whose molecules selectively absorb different wavelengths is a *dye*. Dyes can be applied directly to some materials or used to create pigments that can then be used to make paint. The success of the paint and dye industry over the last 150 years has transformed our world, adding color beyond the imagination of our ancestors.

There are many excellent books on the topic of color in nature. *The Color of Nature* presents a beautifully illustrated overview that is technically accurate, though not very deep. For those looking for more technical detail, I recommend *Light and Color in Nature and Art*, by Williamson and Cummings. Intended as a textbook, it presents an excellent survey of the science of color and its applications. *Color and Light in Nature*, by astronomers Lynch and Livingston, is a detailed description of atmospheric colors of all forms, accompanied by lovely photographs.

This chapter contains a section on each of the mechanisms that create color in nature: emitted light, selective absorption, refraction, interference patterns, and scattering. These sections are followed by a description of the colors in the sky, and a discussion of pigments and dyes, the basis of manufactured color. The chapter concludes with a section on computing with the color of nature, which describes some simple computational models for turning physical color into digital color.

# Light Sources: Emitted Light

Objects give off light because they emit energy in the visible spectrum. The color of the light depends on the distribution of this energy, as described in Chapter 1.

One of the most basic ways of creating light is by applying heat to create a glowing filament or an open flame. Many light sources can be modeled as black-body radiators, which means that the emitted spectral distribution is a strict function of temperature. When heated, such an object first glows red, then orange, yellow, white, and blue-white as the temperature increases. Black-body colors are

Figure 4.
Scattering creates the white clouds and the blue sky. (Copyright Pauline Ts'o.)

Figure 5.
Rainbow colors are caused by dispersion; the raindrops act as prisms. (Copyright Joel Bartlett.)

Figure 6.

A candle emits an orange-yellow light. (Copyright Pauline Ts'o.)

described by their temperature, specified in degrees Kelvin. Candle flames, like the ones in Figure 6, have a relatively low color temperature of around 2,000 K. Incandescent light bulbs, whose light comes from heating tungsten filaments, are blackbody radiators whose color temperature range from 2,500–2,800 K. The bright white light of a carbon arc projector is 5,000–5,500 K; the blue-white light of the star Sirius is 11,000 K.

Figure 7.

The spectrum for a black-body radiator of color temperature 6,500 K, and the equivalent daylight color temperature, D65.

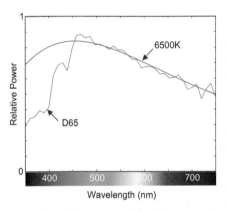

The CIE-defined daylight spectra are also specified by temperature. D65 approximates mid-day, whereas D93 is northern afternoon light. These terms are sometimes used interchangeably with 6,500 K or 9,300 K, but the daylight colors as defined by the CIE are a different spectral distribution than the black-body radiators because they simulate not only the color of the sun (a black-body radiator of around 5,800 K), but also scattered light from the sky. Figure 7 shows the spectrum for a black-body radiator and a daylight color of the same color temperature.

Another common source of light is gas discharge. Adding energy to a gas, such as passing a current through it, can cause electrons to change state and emit energy at specific frequencies. This can translate into visible colors, such as those produced by a sodium vapor street light or a neon sign. Unlike black-body radiators, which produce smooth spectral distributions, gas discharge spectra contain spikes at specific, characteristic frequencies.

Some materials emit visible light when excited with electrons or other energy sources. The phosphors on a CRT are an example of this.

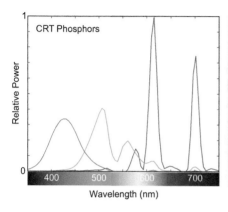

Relative Power

CRT Phosphors

Wavelength (nm)

CRT Display

Figure 8.
CRT phosphors, whose spectra are shown on the left, create the glowing colors of a computer display (right).

All the colors on the display are created by combinations of the phosphor colors (Figure 8). Many light sources use a combination of techniques to create the desired spectral properties and appearance. The inside of fluorescent light bulbs are coated with phosphors that emit a broad spectrum of light, making their output a combination of spikes (from the gas discharge) and the phosphor curve, as shown in Figure 9.

Color temperature can be used as a colorimetric term to describe any light source in the reddish-yellow to blue-white range. Fluorescent lights range in color temperature from warm white (around 3,000 K) to daylight (around 6,500 K). Similarly, the settings on many CRT displays are specified either by color temperature (5,000 K–9,300 K) or as daylight temperatures (D50–D93). Used in this manner, color temperature indicates only the location where the color plots on the chromaticity diagram, not its specific spectral distributions.

It is important to understand when the spectral distribution, such as color temperature, is sufficient to define a color and when the full spectral representation is needed. For example, both a burning

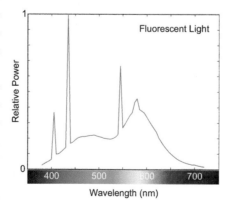

Relative Power

Fluorescent Light

Wavelength (nm)

Figure 9.
A fluorescent light source, showing the smooth spectrum from the white phosphor plus the spikes of the gas discharge.

69

candle and the glowing wire of an incandescent bulb are physical approximations to black-body radiators, so they are spectrally as well as colorimetrically defined by their color temperature. The spectra for a fluorescent light, in contrast, is quite different, so it is only a metameric match to a black-body radiator of the same color temperature. This doesn't matter when viewing the light directly, but if the light is used to illuminate colored objects, the spectral distribution does matter. Lights with different spectra, even if colorimetrically equivalent, can make the same object appear different colors, as will be described further in the next section.

## Surface Colors: Selective Absorption

Take a light and shine it on a colored object such as a leaf, a plastic toy, or a painted surface. The colors you see are defined by the interaction of the light with the surface. Just as a light source is described by its spectral distribution, the color of many surfaces can be defined by a spectral distribution that describes the percentage of light that is reflected, transmitted, or absorbed at each wavelength.

The reflectance spectrum shown in Figure 10 reflects yellow, red, and some green colors and absorbs the rest. Such an object would appear yellow under most white light sources, although its precise color will depend on the spectral distribution of the incident light. Such a spectrum could also describe the color transmitted through a material, such as a piece of transparent yellow plastic. Many objects both reflect and transmit color, such as the flower shown in Figure 11.

The spectrum seen by the eye is the product of the light

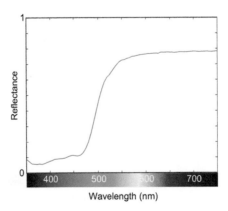

Figure 10.

Reflectance spectrum of the petals of a yellow flower.

70

and the surface spectra, as described in Chapter 1. Most colored objects will be seen under a variety of light sources, creating inevitable shifts in color.

Chromatic adaptation, which was discussed in Chapter 2, generally keeps us from noticing the change in object colors as the lighting changes. The effect becomes visible, however, when trying to match two colored materials, such as two articles of clothing or home furnishings. They may look the same in the store, but not outside or at home. This is a practical example of metamerism. The product of the light and surfaces does not create the same spectra, but creates two spectra that look the same—they are a metameric match. Changing the light produces two new spectra which do not match.

The color of most surfaces can be defined by the simple light-times-reflectance product described above. However, some surfaces absorb light at one wavelength and emit it at another, a property called *fluorescence*. Often, this involves shifting invisible light energy, usually ultra-violet, into the visible spectrum. Many papers and laundry detergents contain fluorescent whiteners, and as a result, are truly "whiter than white." Fluorescence can be used to make colors so vivid they appear to glow.

*Phosphorescence* is similar to fluorescence, but with a time delay. The molecules of the material absorb light energy, but retain it for some time before emitting it as light. Phosphorescence is common in biological systems, and is the mechanism behind glow-in-the dark toys.

Figure 11.

A yellow iris, showing both transmitted and reflected yellow light. (Copyright Pauline Ts'o.)

# Refraction: Bending Light

Light is transmitted through transparent materials, but its path may be altered by refraction. This is the basic principle behind lenses, which bend light rays by passing them through carefully shaped glass. The *refractive index* (or *index of refraction*) of a material defines the angular relationship between reflection and refraction according to Snell's Law

$$n_1 \sin \theta_1 = n_2 \sin \theta_2,$$

where $\theta_1$ is the angle of reflection, $\theta_2$ is the angle of refraction; and $n_1$ and $n_2$ are the refractive indices for the two media, as shown in Figure 12.

Different wavelengths each have a slightly different refractive index, an effect called *dispersion*. This produces the colorful display caused by splitting white light with a prism, as well as the colors of the rainbow. The colored patterns in Figure 13 are caused by dispersion as the light is focused by the irregular surface of the water. The higher the refractive index, the bigger the dispersion—this is why diamonds, with their very high refractive index, sparkle more colorfully than most other materials.

## Interference Patterns: Layers and Waves

Light can be modeled as a wave, as suggested by the application of the term "wavelength" to visible light. When two waves interact, they create *interference patterns*. Think of the patterns created when two rocks are dropped in a pond. As the waves combine, they create wave patterns by amplifying and canceling each other. *Constructive* interference creates larger waves, *destructive* interference creates smaller ones. Figure 14 shows two pairs of waves. When they are in phase, as in Figure 14 (top), their amplitudes sum to create the larger wave, shown as a dashed, red line. When they are perfectly out-of-phase, they sum to zero, as shown in Figure 14 (bottom).

When light reflects off a thin film, such as an oil film on water, it reflects from both the front and back edge of the film. If the thickness of the film is comparable to the wavelength of the light, the two reflected waves will interfere with each other. Figure 15 shows a wave reflecting off the front and back surfaces of a thin film of thickness t. If the wave completes one complete cycle in a distance 2t, it will remain in phase, and the peaks of the two reflected waves will add (constructive interference), as shown in Figure 14 (top). If the wave completes a half a cycle in 2t, the two reflected waves will be perfectly out-of-phase and cancel, as shown in Figure 14 (bottom). The incident light may contain many different wavelengths, in which case the thickness of the film will determine which are amplified and which are cancelled. Color created this way is called *iridescent color*.

The discussion above assumes the material of the film does not change the wavelength of the light, which is only true if the materials have the same refractive index. The more precise specification is: Given a film material whose refractive index is n, the maximum amplification occurs when the wavelength equals 2nt.

The colors created by interference patterns shift as the path difference between the two interfering waves changes. This can happen if the film changes thickness, for example, as a soap bubble dries. Or, it can change with viewing angle, which changes the distance the light travels inside the film with respect to the viewer, as shown in Figure 16. These color shifts are an identifying characteristic of interference colors, such as those seen in the soap bubbles in Figure 17.

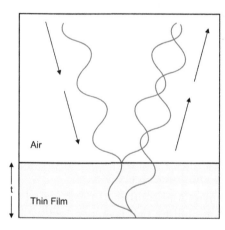

Air

t

Thin Film

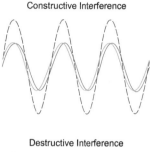

Constructive Interference

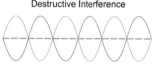

Destructive Interference

Figure 14.

Adding two waves creates constructive interference when the waves are in phase, and destructive interference when they are out-of-phase. In both examples, the blue and green waves sum to the red, dashed one.

Figure 15.

How thin films create interference colors. The incoming light reflects off both the front and the back of the thin film. The resulting phase shift creates interference patterns.

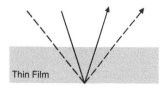

Thin Film

Figure 16.

The path of the light, and hence, the distance traveled inside the thin film, changes with angle.

Figure 18.
The fine structure of peacock feathers creates interference colors that change with viewing angle.

Fine structures whose spacing is close to the wavelength of light can also create interference patterns. Peacock feathers, such as those shown in Figure 18, are a classic example. The color changes as the viewing angle changes, as can be seen by the slightly different colors in the different feathers. Look especially at the color of the three blue "eyes." Peacock feathers also contain pigments—their colors are a combination of absorption and interference effects.

The flashing colors seen in opals and mother-of-pearl are also caused by interference, which is created by fine layers embedded in the material. Such colors are often called *opalescent* colors.

# Scattering: Small Particles

When light strikes particles suspended in a medium, such as smoke or dust in the air, it scatters. The way scattering creates color depends on the particle size: larger particles reflect and absorb like surfaces. The color produced is a function of the reflected light, which is scattered back towards the observer—this is how pigments produce color. Dense clouds of smaller particles near the wavelength of light can cause constructive and destructive interference. Iridescent colors in thin, high clouds seen near the sun are interference patterns caused by water or ice droplets.

*Rayleigh scattering* applies to particles that are smaller than the wavelength of light. Scattering from such particles depends strongly on wavelength, such that blue light is scattered more than red. The intensity of the scattered radiation varies as $\lambda^{-4}$, making the intensity of scattered blue light ($\lambda = 450$ nm) six times greater than the intensity of scattered red light ($\lambda = 700$ nm).

Rayleigh scattering is responsible for our blue sky and for the color of blue eyes (Figure 19). It accounts for the bluish tint in smoke, and the bluish sheen in black hair. The streaks of blue color seen in the glacier in Figure 20 are caused by Rayleigh scattering from very fine particles trapped in the ice. The skin of many tropical fish, frogs, and lizards contains a layer of fine particles over an absorptive layer that creates vivid blue colors by Rayleigh scattering—blue light is scattered back, the rest is absorbed by the layer beneath. If these layers are overlaid by a layer of transparent yellow pigment, the resulting color will be bright green.

Figure 19.
Blue eyes contain no pigment. The color is caused by Rayleigh scattering from protein particles suspended in the iris.

Figure 20.
The blue in the glacial ice is caused by Rayleigh scattering from dust trapped in the ice. (Copyright Pauline Ts'o.)

# Atmospherics: Sky, Rainbows, Clouds

The colors in the sky come primarily from scattering—white clouds and the blue sky are both scattering effects, caused by sunlight striking small particles in the atmosphere. Sky blue is primarily Rayleigh scattering caused by nitrogen molecules; white clouds contain larger particles of water or ice that scatter all wavelengths uniformly. The red colors of sunsets (Figure 21) are caused by viewing the sun's light through a thick layer of air. The blue light has been scattered away, leaving the red colors of the spectrum. These red colors are

Figure 21.

The red sky of a Los Angeles sunset. (Copyright Pauline Ts'o.)

enhanced by absorption and scattering caused by larger particles in the air, which is why smoke or smog enhances the sunset colors.

The colors of a rainbow (Figure 22) are a spectacular display of dispersion, where water droplets split the sun's rays like a prism. Sunlight is focused inside a spherical drop of water, and reflects off of the back surface. Because different wavelengths refract at slightly different angles, different colors exit at slightly different positions, causing the rainbow. The result is a cone of colored light whose interior angle is 42°, as shown in Figure 23. Each droplet creates its own cone. An observer sees an aggregation of the colors from many droplets. A secondary bow, caused by light that follows a path with two reflections inside the water droplet, can appear outside of the primary rain-

Figure 22.

Double rainbow against a blue sky, Big Island, Hawaii. (Copyright Mary Orr.)

Figure 23.

Each water droplet creates a rainbow by focusing light inside of it. The result is a cone of colored light, whose interior angle is 42°. (Copyright The Exploratium, www.exploratorium.com.)

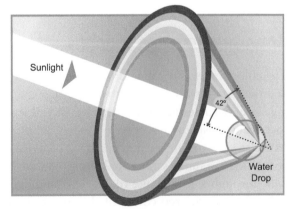

bow at an angle of approximately 51°. Its colors are reversed due to the additional reflection, and dimmer, because of the light lost on the extra bounce. This mechanism produces rainbow-like displays from any clear droplets, including dew drops, ocean spray, or a garden hose. The finer and more uniform the droplets, the brighter the rainbow. Figure 24 shows a section of a double rainbow with particularly intense colors.

Figure 24.
Double rainbow with especially intense colors. (Copyright Joel Bartlett.)

Halos, glories, sundogs, and coronas are other rainbow-colored atmospheric effects. Some are created by refraction, like rainbows. Others are interference colors—for details, including more detail on the complex optics of rainbow construction, consult Lynch and Livingston's book.

The aurora borealis, or northern lights, are caused by ionized gasses in the atmosphere discharging colored light. They are stimulated by radiation from solar flares, and are usually visible only near the poles.

# Pigments and Paints

A pigment is a colored material ground into a fine powder. The first pigments were created from naturally occurring mineral compounds such as carbon (black), iron oxides (reds and browns), sulphur compounds (yellows and oranges), copper compounds (blues and greens) and calcium carbonate (white). Such pigments supply the colors in the native American pottery in Figure 25.

From antiquity, people have sought new pigments and efficient processes to create them. Early processed pigments included

vermilion (mercury sulfide), verdigris (copper acetate), ultramarine (extracted from lapis lazuli) and azure (from azurite, a copper carbonate). Indigo and madder are pigments extracted from plants; in general, organic pigments are less stable than inorganic compounds. Organic dyes, such as crimson, can be combined with colorless particles (usually alum) to produce a *lake* pigment. Today, both natural and artificially produced pigments are available in a wide range of colors.

Pigments are mixed with water, oil, egg (tempera), or some other medium to create paints that can be applied to a surface. Some methods, such as water color, simply leave the pigment stuck to the surface (water color paints contain a small amount of glue to help this process). Others, such as oil or tempera paints, dry to form a thin layer containing a suspension of pigments. The resulting color is created by a complex blend of scattering and absorption from these particles. The vividness and opacity of the color depends not only on the surface properties of the particles, but on their size, the thickness of the paint layers, and the optical properties of the medium.

Figure 26 shows a layer of particles suspended in a medium, such as an oil paint, and several possible light paths. Incoming light rays can reflect off the surface of the medium (1) or off a pigment particle (2). In these cases, no absorption takes place and the light remains uncolored. If the light passes into the pigment, some of the wavelengths are absorbed, imparting a color. Light can either scatter

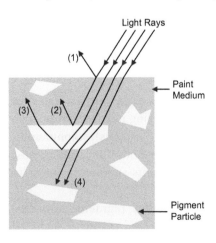

Light Rays
(1)
(3) (2)
(4)
Paint Medium
Pigment Particle

Figure 26.
Light shining on pigments suspended in a paint can take many paths. The more the light interacts with the pigment particles, the more intensely colored it becomes.

78

inside the pigment (3) or pass through it (4). Light remaining inside the paint layer can interact with other pigment particles or strike the underlying surface, which is often treated to efficiently reflect light back. The color produced is the additive sum of all the light rays scattering out of the surface of the paint.

The degree of scattering depends on the difference between the refractive index of the particles and the medium, as well as the density of the particles. The higher the relative refractive index, the more the light scatters. White pigments, such as titanium dioxide, are colorless particles with a high refractive index—they absorb almost no light, creating a bright, white paint.

Light is absorbed by molecules in the pigment. The longer the light path through the pigment, the more opportunity for the unwanted wavelengths to be absorbed. Big particles with a low refractive index will absorb well but will scatter poorly, so more light will reach the underlying layer. Carried to extremes, the paint would be grainy, and would not be very opaque. A good paint needs both absorption and scattering to create a bright, uniform color.

The opacity of a paint depends on how much light is scattered by the paint and how much is reflected from the layer beneath. Grinding pigment into small particles will give a paint that is more opaque but less saturated than the same volume of pigment coarsely ground.

Transparency can be desirable, as in pigmented pottery glazes. Certain painting techniques, as in the oil paintings produced by the Dutch masters, carefully build up layers of overlapping color, giving great depth and richness to the painting. To achieve complete opacity, a surface is usually primed with a white, opaque undercoat, or coated with several layers of paint.

Many colored materials contain pigments. India ink is carbon black in a water-based medium. Pigments create colored pencils, crayons, and pastels (inks in markers, however, are usually dyes). Cosmetics are pigments in oils. Plastics are colored with pigments, as are colored glazes on pottery and certain forms of stained glass.

Pigments occur naturally in skin and hair. The brown pigment melanin gives humans our skin tones—the higher the concentration, the darker. Beta carotene is a common pigment responsible for many of the red-orange colors found in living beings, from carrot orange to flamingo pink. Transparent yellow pigments overlay the skin of many reptiles and amphibians, tinting the colors underneath.

## Dyes: Cascading Filters

Dyes are chemicals, often organic compounds, whose molecules selectively absorb different wavelengths of light. Therefore, all of their color properties come from absorption—there is no scattering as in pigments. Dyes dissolved in clear gelatin or plastic create colored filters, such as those used in stage lighting or photo processing. Dyes dissolved in solution color fabrics. Layered dyes are the basis for the subtractive color reproduction processes—photography and printing, as will be discussed further in Chapter 8.

The earliest dyes were derived from plants and animals. Solutions made by crushing or soaking berries, leaves, or roots have been used to dye cloth for millennia. Brilliant red dye comes from the cochineal insect. Tyrolean purple is extracted from shellfish.

Around 1850, William Perkin extracted the first commercially successful synthetic dye from coal tar, a sticky black substance that is rich in organic molecules. This color, called mauve, was a bright purple that rivaled expensive Tyrolean purple. Further research by Perkin and others created a rainbow of colors, advanced the science of organic chemistry, and formed the foundation of the modern chemical industry.

Predicting the color of a dye is much simpler than for a pigment. The dye solution

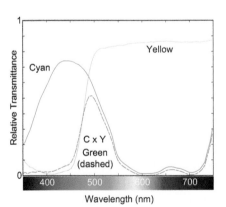

Figure 27.

Printing inks are dyes whose spectra multiply to create the final color. Cyan × Yellow = Green.

has a characteristic absorption curve that filters color from a white light source. Combining two dyes simply multiplies their curves. This is always true for layered filters (modulo loss due to internal reflectance), and may be true for combined solutions, assuming no chemical interactions. This is illustrated in Figure 27, which shows a combination of the spectra for two printing inks, cyan and yellow, to create green. While it is often stated when describing printing colors that cyan + yellow = green, spectrally, it is cyan × yellow = green.

There are two basic challenges to making a successful dye: making it colorfast, and making it stick to whatever you are trying to color (typically fibers or fabric). *Colorfast* means that the color lasts, even when exposed to light. Many dyes fade quickly, especially if exposed to sunlight. To make a dye adhere to a material, there needs to be some form of electrochemical bond between the dye and its substrate. Often, an additional chemical called a *mordant* is applied to make the dye adhere. Some dyes react chemically with the material being dyed. An ancient example is henna, which reacts with the protein in hair or skin. Modern *reactive dyes* can create very bright, permanent colors—the vivid colors seen in cotton clothing nowadays, such as those shown in Figure 28, are typically created with reactive dyes.

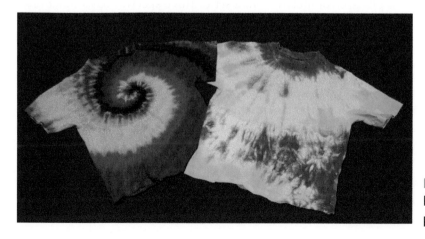

Figure 28.
Brightly colored dyes create the patterns in these tie-dyed shirts.

# Computing the Colors of Nature

Digital simulations of light and color are used in many fields. Such simulations are used to compute the color of lights and objects for the digital images created in the field of computer graphics, which is discussed in Chapter 10. This section provides an introduction and some resources for those interested in computing with physical color.

Lights, and colors created by selective absorption (including dyes), can be represented as spectra and computations on spectra. Most spectral data are samples taken with a spectroradiometer or spectrophotometer. Typical high-resolution samples are taken every 2 nm, or even every 1 nm. Such resolution is important if the spectrum has narrow spikes such as those produced by gas discharge. Most surface reflectance functions are smooth, so a 10–20 nm spacing is often sufficient for these measurements. There is also variation in the range of wavelengths sampled, with 350–750 nm being common, but not universal. Sampled spectra may contain absolute power values, or normalized values. There are different units for spectral measurements, and different conventions for normalization, so care is needed when using spectra from a variety of sources.

Spectral data can be represented as weighted basis functions, which means that a small set of spectral distribution curves can be used to compute a family of related curves. The CIE daylight spectra, for example, are computed from a set of three spectral basis functions. The general form of the computation is

$$s(\lambda) = w_0 B_0(\lambda) + w_1 B_1(\lambda) + \ldots + w_n B_n(\lambda),$$

where $B(\lambda)$ are the basis functions, w are the weights, and $s(\lambda)$ is the resulting spectral distribution. Basis functions can be created from any set of data using a mathematical technique called *singular value decomposition*, though the accuracy will depend on the coherence of the dataset and the number of basis terms used. Maloney and Wandell applied these principles to the description of surfaces, as illustrated in Figure 29. The graph in Figure 29 shows the first four basis functions for the colors in the Macbeth Color Checker, which is a test chart used by photographers (see Chapter 6). The images on

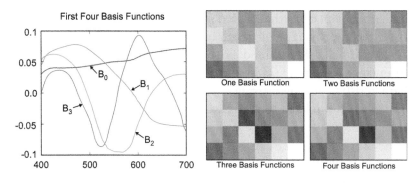

First Four Basis Functions

One Basis Function    Two Basis Functions

Three Basis Functions    Four Basis Functions

Figure 29.
The surface reflectances of the Macbeth ColorChecker can be decomposed into basis functions. The plot shows the first four basis functions; the images show the result of reconstructing the colors using progressively more basis functions. (Data and renderings created by Brian Wandell.)

the right show the results of rendering the colors using progressively more basis functions. This example is discussed in further detail in Chapter 9 of Wandell's *Foundations of Vision*.

Spectral data is often shared on the web. Most universities and research institutions doing work in color publish their data on the web. The Munsell Color Science Laboratory at the Rochester Institute of Technology is one good example. In addition, individuals with an enthusiasm for color will often create websites containing spectral data. Web-searching will turn up many resources.

Black-body radiation is used in many fields, from engineering to astrophysics. The equations to compute a spectrum from a color temperature are complex, but there are many implementations on the web, both as software packages and as Java applets. Bruce Lindbloom has such a calculator on his website (www.brucelindbloom.com). Other resources can be found on the web as well.

Colors from interference or dispersion require modeling the geometric interaction of light rays with surface structures. Ray tracing is the technique most commonly used for this. Ray tracing originated in optics, but is now a fundamental part of computer graphics rendering (see Chapter 10). POV-ray is a well-know open source graphics package for ray tracing (www.povray.org). Web-searching returns references to optical simulators as well.

The color of paint is a complex blend of scattering and absorption, but the Kubelka-Munk model provides a simplified method for simulating paint color that is used in many fields. This model is often applied to approximate the result of mixing paints or inks.

The Kubelka-Munk model depends on two empirically derived parameters, $K(\lambda)$ and $S(\lambda)$, that define the absorption and scattering characteristics of a pigment in a particular medium. In its simplest form, these are related to the reflectance, $R(\lambda)$, as

$$K/S = (1 - R)^2/2R,$$

where R is the reflectance for a paint layer so thick the substrate does not show through. This can be applied to a paint mixture by defining the K/S of a mixture as the sum of the individual pigment values. For a mixture of n pigments,

$$(K/S)_{mixture} = (c_1K_1 + c_2K_2 \dots c_nK_n)/(c_1S_1 + c_2S_2 \dots c_nS_n),$$

where the $c_n$ are the concentrations for each pigment in the mixture.

Figure 30.

Computer graphics image demonstrating simulated paint using the Kubelka-Munk model. (Image by Chet Haase and Gary Meyer, copyright ACM, Inc. Included here by permission.)

Typically, these K/S values are determined by first measuring R for a white paint and setting S to uniform scattering ($S(\lambda) = 1$). This defines K and S for the white paint. Measuring both the pigment and a known mixture of the pigment with white will produce two K/S values. This creates enough variables and equations to solve for K and S for the pigment.

The Kubelka-Munk equations are used in various industries to predict the color of paint or ink mixtures, and in computer graphics to predict mixtures and the appearance of simulated paint. Figure 30 is an early computer graphics rendering of paint mixture from an article by Haase and Meyer. The mathematical analysis for using the Kubelka-Munk model is presented in more detail in their article and also in the book by Williamson and Cummings.

Reference.

Chet S. Haase and Gary W. Meyer. "Modeling Pigmented Materials for Realistic Image Synthesis." *ACM Transactions on Graphics* 11 (1992), 305–335.

# Summary

The colors in nature result from light striking objects and can either be defined simply as the products of spectra or involve interactions defined by the wave nature of light. While most colors come from selective absorption, reflection, and transmission of light sources, some of the most spectacular colors are created by fracturing light into its colored components, either by dispersion, refraction, or selective scattering.

The colors in nature stimulate the eye as spectral distributions, but the appearance of objects is a complex mix of surface properties: shiny surfaces have highlights, rough objects have texture. A surface may consist of several reflective layers, such as a waxed car body, or scatter light within it, as in a block of marble. It may be a complex blend of surfaces, such as fur or hair. Rarely is an object color well-represented by a single spectral distribution.

Digital models for physical color range from sampled spectra to the sophisticated industrial models used for designing and predicting manufactured colors. Many paint stores, for example, will match any colored object using modeling software, though an experienced colorist can often do equally well by eye. I once heard of a color-mixing model to help solve the problem of wasted ink in a large gravure printing press—given the volume of the current color left in the press and the desired color, it computed what color of ink to add to achieve the desired result. Most such industrial models, however, are highly proprietary.

Digital models for color are used in computer graphics to simulate the colors of nature: some are accurate simulations of these physical models while others are highly simplified. Color in computer graphics will be discussed in Chapter 10.

# 5
# Color Reproduction

Take a camera, point it at a scene full of lights and objects, then snap a picture. The resulting photograph is a color reproduction of the original scene. In this particular example, light is reflected from the objects in a room and focused on the film, which chemically reacts in response to the light. Developing the film produces a picture that, hopefully, looks very much like the original scene. The quality of the reproduction depends on the camera, the film, the processing, and the skill of the photographer, whose knowledge of the characteristics and limitations of film as a medium are critical to achieving a good result. This is the first chapter on the field of color reproduction. It provides an overview and defines several key principles. Subsequent chapters will provide further detail about specific aspects of color reproduction such as image capture, additive and subtractive color systems, and digital color management systems.

# Introduction

Color reproduction traditionally means reproducing colored images of the natural world. It has been defined by the different technologies used in the traditional color reproduction industries of television, printing and photography, and cinematography. Over the last couple of decades, the application of digital technology has profoundly changed these industries. Especially in the graphic arts, color reproduction is moving from specialized, end-to-end systems to general-purpose capture, encoding, and rendering of digitally defined images.

The primary goal in traditional image reproduction is to create output that looks like the original. It is worth being specific at this point about what "looks like" means. One ideal goal would be that the reproduction is indistinguishable from the original. This is clearly only possible when the original is very similar to the copy, such as a duplicate print or slide. Even converting from a slide to a photographic print requires compromises in contrast and colorfulness because photographic paper can produce a more limited set of colors than slide film.

Differences between the original and the reproduction are an unavoidable part of color reproduction; each medium has its own characteristics, and no medium can match all of what the eye can see in nature. Color reproduction, therefore, is both an art and a science. Most color reproduction industries must balance time and cost against quality, adding a commercial constraint as well as the technical and aesthetic ones.

Traditional color image reproduction is a linear process, as illustrated in Figure 1. It begins with lights and objects, whose colors are

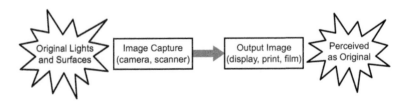

Figure 1.
The traditional, linear process of color reproduction.

first captured, then output as an image in some two-dimensional, pictorial form. The colors in the resulting image, while created quite differently, should evoke the appearance of the original scene. Each of the traditional color reproduction industries has its own linear pipeline whose components and processes are tuned specifically for that industry.

In digital color reproduction, this linear pipeline is replaced by a star-shaped model, as shown in Figure 2. There are many different sources of images, and many different paths for output. While it used to be sufficient to capture images for specific media, the same image may now be printed, projected, or published on the web. Conversely, images captured or created in many different ways may be combined for a single print, web page, or slide, as well as stored or edited.

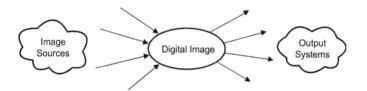

Figure 2.

The star-shaped model introduced by digital color reproduction.

To support this new star-shaped model, we need some way to specify color that is independent of any particular color reproduction technology. Encoding colors and characterizing color technology with respect to perceptual values such as CIE XYZ or CIELAB provides a *device-independent color representation*. Such representations are a key component of *color management systems*, which are specifically designed to solve the problems associated with using a star-shaped image reproduction model. Device-independent color representations also provide a scientific specification for digital color in a wide variety of applications.

The discussion of color reproduction in this book will focus on device-independent color reproduction, as digital color created the star-shaped model. Once you can manipulate the pixels with a computer, they have escaped from the traditional linear pipeline.

The classic reference for the topic of color reproduction is Robert Hunt's *Reproduction of Colour In Photography, Printing &*

*Television.* First published in 1957, the fifth edition was published in 1996. The topic of device-independent color reproduction has no equivalent text, but a recent edited collection of articles, *Colour Engineering: Achieving Device Independent Colour,* by Phil Green and Lindsay MacDonald, provides a good overview of the field.

This chapter is the first of five on color image reproduction. It provides an introduction and focuses on the principles applied to creating a good reproduction independently of the technology involved. It also shows, by a sequence of examples, how color reproduction has been traditionally practiced, and how traditional practice has changed with the introduction of digital color. The following three chapters describe each of the primary fields of color reproduction: image capture, additive color systems, and subtractive color systems. In each case, the emphasis will be their use in digital device-independent color reproduction systems. The final chapter of the five color reproduction chapters describes color management systems, which need to combine device-independent color representations with the principles and craft of color reproduction systems.

## Digital Color Reproduction

The process of digital color reproduction is shown in Figure 3. An image is captured, either from a natural scene or scanned from a photograph or other print, and encoded as digital pixels. These pixels are made visible by sending them to an image output system, such as a display, printer, or film recorder, to create an image in some physical form. There are color transformations at each step in this process, which must be controlled to create a good color reproduction.

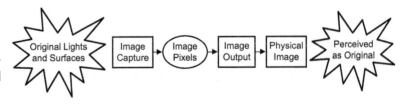

Figure 3.

The process of digital color reproduction, where images are described as digital pixels.

The original colors are spectral distributions created by the interaction between lights and objects, or glowing directly from light emitting sources, as described in Chapter 4. Looking back to the discussion of color measurement and encoding in Chapter 1, we know that for a fixed set of lighting and viewing conditions, we can encode the colors reflected from a scene as three numbers that represent roughly the red, green, and blue components of the colors seen. *Image capture* is the process of capturing and encoding these values for every point across the image—the light from the scene is split into the familiar RGB image pixels used in digital representations of images, as illustrated in Figure 4. Further detail about image capture will be provided in Chapter 6.

Image output systems convert numeric pixels to some physical form such as a display, print, or film. There are two types of processes used by image output systems. *Additive color*, the process used on light-producing media such as monitors or digital projectors, adds together red, green, and blue light to create a full-color image, as shown in Figure 5. *Subtractive color*, used in film and print, consists of dye layers (cyan, magenta, and yellow) that subtract the red, green, and blue components from a white light source. Each filter modulates only the red, green, or blue separation, as shown in Figure 6. Additive reproduction adds three separate colored sources; subtractive reproduction modulates a single white light through three overlapping filters. Further detail about additive and subtractive reproduction will be provided in Chapters 7 and 8.

The *gamut* of a particular color reproduction device (printer, display, etc.) is the set of all possible colors that can be created using that device. Different devices have different gamuts, and all device gamuts are much smaller than the full set of colors in nature. For example, an outdoor scene may have a contrast ratio of more than 1000:1, whereas the ratio of white to black on most physical image media is less than 300:1. While most color media can reproduce colors of all hues, how bright and how saturated the colors can be at different brightness levels is often very different. For example, monitors can produce bright, highly saturated blues, whereas similarly saturated blues in print are very dark. Print and film, however,

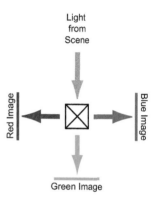

Figure 4.

Image capture: spectral representations of light and object colors are converted to red, green, and blue pixels.

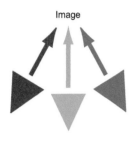

Additive color reproduction: Red, green, and blue light is combined to create an image.

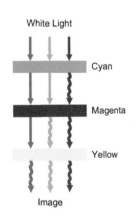

Subtractive color reproduction: White light is filtered to create an image. The wavy lines show each color being modulated in turn by the cyan, magenta, and yellow layers.

can create vivid, golden yellows that are simply not available on most displays. Gamut limitations are one of the key reasons for frustration with color image reproduction.

The process of image reproduction is one of continuous approximation and compromise. The best results are achieved with finely tuned end-to-end systems, where even the original is adapted to the reproduction process. For example, television stages are lit to reduce the overall contrast, making the scene more compatible with television display systems. A graphic artist creating a printed brochure will constrain design colors to lie within the print gamut, traditionally by picking them from printed samples.

In the new, computer-mediated digital domain, we do not have a single path from capture to output. Figure 7 shows the many possible color reproduction paths in a digital color system. An image is captured either directly with a digital camera, or indirectly by scanning a photograph or print. Digital images can also be created directly, using computer graphics to render synthesized lights and surfaces or by using digital painting and drawing systems, or ultimately, by any program designed to generate pixels. These pixels can then be edited and combined with other images and colors. The result can be output to a variety of digital media (printers, projectors, film recorders, video recorders, web pages) or stored for use at some other time or place.

This star-shaped process is much more difficult to manage and control than the traditional linear one. Gone is the ability to tune the input to match the output—the image must stand independently. And while traditional image reproduction systems can be tuned to the reproduction of "natural" scenes, digital images often include colors and shading never seen in nature. Engineering decisions made for natural scenes may fail dramatically for synthetic ones.

Device-independent color and color management systems (which will be discussed more fully in Chapter 9) were invented in response to the demands of digital image reproduction. Images and output systems are *characterized* to a perceptual standard such as CIE XYZ or CIELAB, which provides a concise specification of image and device gamuts and an algorithmic way to transform from the de-

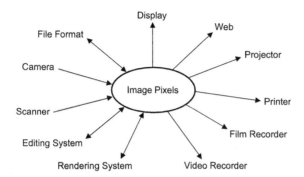

Figure 7.
Digital color reproduction maps many
different inputs to many different
outputs.

vice-independent color space to a device-specific one. This provides a metric foundation for describing digital color. Unfortunately, it is not the complete solution to the problem of digital color reproduction. Because of gamut limitations, and because image appearance is more complex than color measurement principles (as discussed in Chapter 2), we cannot simply match the image color point by point to achieve a good reproduction. The colors must be transformed, and the best way to perform these transformations is not well-defined. As in all image reproduction, some part of the answer will always be application-specific.

The rest of this chapter begins with the basic principles and processes that define and control quality in color image reproduction. It concludes with three examples of color reproduction systems, which illustrate these principles.

# Principles of Color Image Reproduction

There are several basic principles used to describe good quality color image reproduction. While their order of precedence is sometimes debated, the common elements are:

Correct mapping of critical reference colors such as sky, foliage, and skin tones. This may not mean any sort of exact match, but simply that the reproduced color is not clearly wrong.

For example, almost any shade of blue will produce a satisfactory sky; even shades approaching purple might be acceptable, but green is clearly wrong.

Correct mapping of white and the neutral colors that constitute the *gray axis* or *neutral axis*, which runs from black to white. These colors should appear neutral, otherwise the image will have an unpleasant *color cast*, or overall color tint.

Control of the tone reproduction, which is the mapping of the overall brightness and contrast. Image reproduction often involves compressing the tone scale. The goal is to reproduce, as best as possible, detail at all levels of brightness throughout the image while maintaining the correct overall appearance of the image.

Control of the overall colorfulness, so that the image doesn't appear either washed out or gaudy.

Control of sharpness, texture, and other visual artifacts that contribute to image appearance. This is particularly important for printing, where the halftoning patterns may be visible enough to mask the detail in the image.

All other color adjustments.

As well as describing goals for an image reproduction process, these principles provide a vocabulary for describing and adjusting image appearance. Color image reproduction often involves a great deal of color image editing in programs such as Adobe Photoshop. These adjustments are often necessary to correct flaws in the image, such as a color cast introduced by the image capture system, but can also be used to adapt the images to the reproduction process as needed.

The process of color image reproduction involves many complex steps, all of which affect the final color. *Process control* is used to maintain consistency throughout the system. While process control will affect all of the aesthetic factors described above, a reliable and controllable image reproduction system is partitioned to separate process control, which monitors the basic operation of the system, from image-specific aesthetic transformations.

The next section discusses tone mapping and color balance, which are the primary adjustments used in image manipulation, and also in the control of image reproduction processes and systems.

# Tone Mapping and Color Balance

*Tone mapping* defines the way pixels (or other input/output values) map from dark to light. *Color balance* describes the relative weights of the color primaries, either red, green, and blue or cyan, magenta, and yellow. Tone mapping affects brightness and contrast, which impact both the overall appearance and how much detail is visible throughout the image. Color balance affects the hue and the relative brightness and saturation of colors. These concepts are common to all image reproduction industries, though the terminology varies.

Controlling tone mapping is a key part of any color reproduction process. Tone-mapping curves are used to describe various mappings from input to output lightness values. Such curves may map pixels to pixels, as in image editing programs, or they may represent physical measurements, such as input intensity to output intensity, as in the brightness scales described in Chapter 3.

Figure 8 shows the effect of changing the tone-mapping curve for a grayscale image and for a color image. The curve applied is shown next to the images. For the color image, the curve is applied to each of red, green, and blue. This has the effect of changing not only the brightness and contrast, but also the saturation of the colors. The tone-mapping curves are pixel-to-pixel mappings. A pixel value of 0 is black, and a pixel value of 1.0 (255 for byte values) is white. Figure 8(a) shows the original image, with a linear tone-mapping curve. Figure 8(b) shows a curve that increases contrast by making only the dark pixels (shadows) darker and the bright pixels (highlights) brighter while leaving the middle values (midtones) relatively unchanged. Figure 8(c) shows a curve that decreases brightness by making most of the pixels darker, while Figure 8(d) shows its inverse, which makes the pixels lighter.

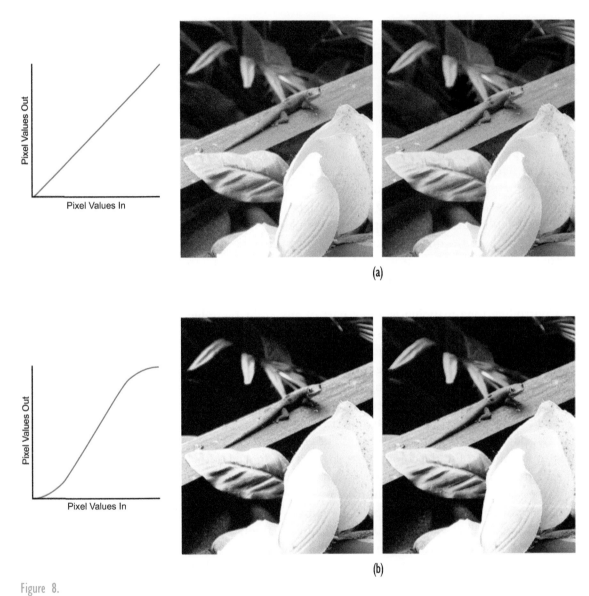

(a)

(b)

Figure 8.
The effect of tone mapping on image appearance. The curves show the mappings. (a) Original image; (b) a contrast-enhancing curve.

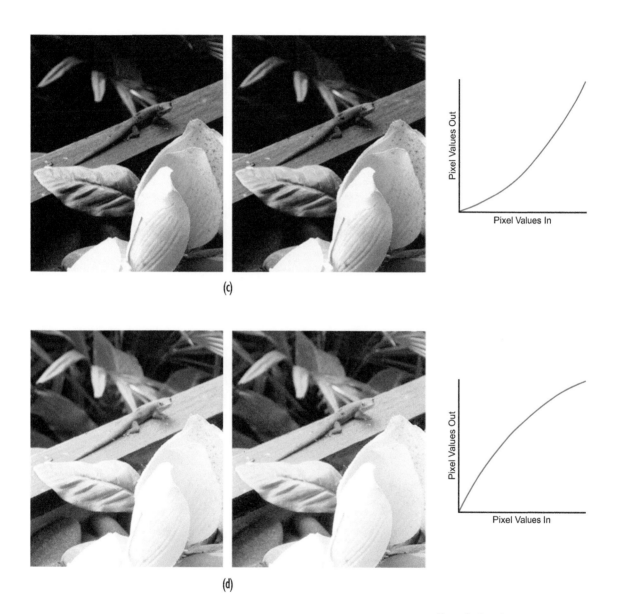

(c)

(d)

Figure 8. (cont.)
(c) A curve that makes most colors darker; (d) a curve that makes most colors lighter.

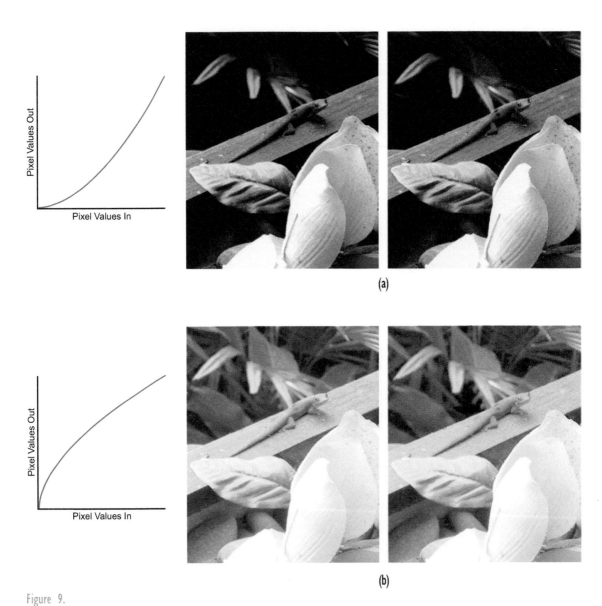

(a)

(b)

Figure 9.
Using a gamma function for tone mapping, where the pixels are raised to the $1/\gamma$ power. (a) $\gamma = 0.56$; (b) $\gamma = 1.8$.

Many image editing programs include more constrained methods for adjusting the tone mapping than editing a curve. Often, they are based on some power function of the form $P_{out} = P_{in}^{1/\gamma}$, where the user is asked to specify the value for gamma: larger gamma values make the image brighter; smaller values make it darker. Figure 9 shows the effect of applying such a function.

Applying different curves to the red, green, and blue components of the image will change the color balance. Figure 10 shows various reproductions of the central image, each with a different color balance. In this example, the variations are fairly large to create a good illustration.

Adjusting color balance is a more complex problem than changing the tone mapping because it involves visualizing how the three primary colors combine throughout the color space. Even for the experienced, it can be difficult to analyze and correct color balance problems without some reference color or image. Furthermore, many color balance problems can mimic the type of lighting changes that trigger chromatic adaptation, which was discussed in Chapter 2. Again, a reference is needed to keep the visual system from adapting to the tinted picture.

To get the best reproduction, the image should have pixels throughout the entire tone range, as it takes full advantage of the dynamic range of the medium. This is visualized by plotting the image histogram, which shows the number of pixels at each pixel value. Figure 11 shows the histogram for the gecko picture in Figure 8(a), which has a good distribution across the entire pixel range. If the image does not have such a distribution, stretching the histogram to map the darkest pixels to black and the brightest ones to white is typically the first step to creating a good tone mapping. Figure 12 shows an example of the type of improvement that can be achieved with this step. The jagged appearance of the histogram is typical when this type of correction is applied to pixel values.

The concept of tone mapping and color balance can be applied to image reproduction systems. *Tone reproduction* is the application of tone mapping to an entire color reproduction process. In this case, the tone-mapping curves are often called *tone-reproduction curves*

Figure 10.
Color variations created by systematically changing the relative amounts of red, green, and blue (or cyan, magenta, and yellow) in the image. The original image is in the center.

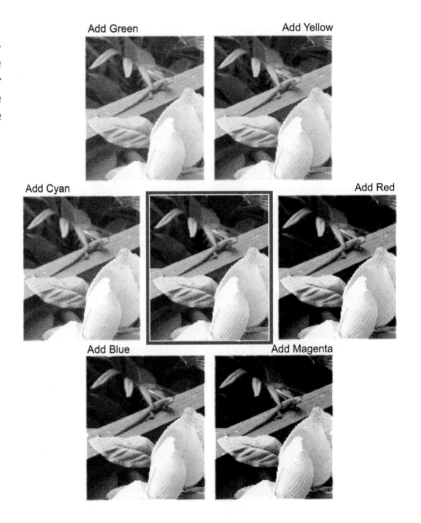

Figure 10.
Color variations created by systematically changing the relative amounts of red, green, and blue (or cyan, magenta, and yellow) in the image. The original image is in the center.

or TRCs, and define the relationship between "original" and "reproduced" values, measured as intensity, reflectance, or density. Color balancing a color reproduction system ensures that neutral colors remain neutral, and that hues are faithfully reproduced. Measuring color balance, however, is more com-

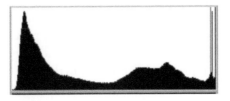

Figure 11.
The histogram for the gecko picture in Figure 8(a). The spike near the right end is the white flower.

100

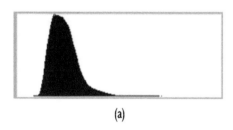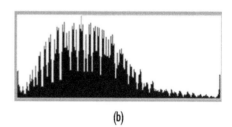

(a)                                                    (b)

plicated than measuring tone reproduction, especially if the color representation change from capture to output.

Any input or output process can and should be adjusted to create an optimal, or at least, predictable, tone mapping and color balance. Examples include the mapping from input signals to displayed light or the transformation from CMYK pixels to printed inks on paper, or the capture of a scanned image into RGB pixels. This is a fundamental part of process control for color image reproduction, as will be described in the next section.

Figure 12.
(a) A picture with few bright pixels, as can be seen from its histogram; (b) redistributing the pixels to fill the brightness range dramatically improves the image appearance.

# Process Control versus Image Adjustment

Converting pixels to physical color is an electromechanical process that has many variables. Even a simple, self-contained desktop printer can accept different papers, which affects the color, and often has a hidden wealth of controls that affect the brightness, contrast, and color balance. A commercial offset printing process includes scanning, plate production, page composition, and finally, printing on a complex electromechanical system controlled by an operator who adjusts the color even as the job runs. Film is manufactured and developed, cameras have exposure controls, displays have brightness and contrast knobs ... the list goes on and on.

Process control ensures that the basic reproduction system is functioning within some standard specification. It usually involves carefully designed test patterns that both provide visual feedback and can be easily measured to obtain a more accurate check on the process. The NTSC color bars, which appear at the start of many video tapes, are an example. Visually, it is easy to see if a color is missing, or if the bars are smeared or otherwise distorted. The bars also produce a standard electronic signal that can be hooked to an instrument called a *waveform monitor* for a more rigorous check of the video signal.

For any reliable color reproduction system, it is important to draw a line between those parts of the process that should remain fixed in one or more standard configurations, and the place in the process where the aesthetic, image-specific adjustments will occur. Otherwise, it is impossible to get consistent, repeatable results—there are just too many variables.

Figure 13 shows a process diagram for the output part of a color reproduction process such as a color printer or display. The "Image-specific Transformation" prepares the original image for reproduction. This may involve fixing problems in the original such as a yellow tint caused by indoor lighting in a photograph, or adjusting the tone mapping for that specific image, or even adjusting specific portions of an image, to correct difficult colors or to enhance detail in dark shadows. The "Process-specific Transformation" takes this corrected image into the color space of the output process. This may be an additive to subtractive color conversion, as in printing, or simply a remapping from one RGB color space to another.

There are two feedback loops in the process. The one marked "Process Control" is independent of any specific image. It is used to maintain

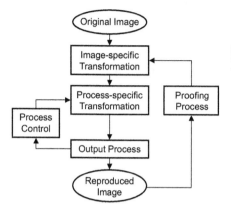

Figure 13.
Process control versus image adjustment. It is important to control them separately.

the tone mapping and color balance of the output process, primarily through the use of standard test patterns and measurements. The one marked "Proofing Process" provides feedback for a specific image. Changes based on this process should be made by adjusting the image, or adjusting parameters to the process-specific transformation.

The more images that flow through the pipeline, the more likely this disciplined separation between process-specific and image-specific transformations will be maintained. A large graphic arts or film production house will have layers of process-specific and job-specific controls. Think, for example, of all the images that must be generated and processed for a movie with digital special effects. Reducing and normalizing all the ways colors can be adjusted is crucial to creating a quality product. In contrast, a small printer reproducing a painting for an artist can tweak any step in the process to create the desired result.

The next sections give three examples of color reproduction systems to demonstrate these principles.

# Broadcast Television Example

In broadcast television, a video camera captures the scene and encodes it as a sequence of images. This encoded representation is displayed on a television, which creates colors by exciting red, green, and blue phosphors with an electron gun. We perceive the light emitting from the front of the picture tube as a sequence of images that reproduce the original scene. Television is an example of additive color reproduction, which involves carefully mapping one RGB color space into another. Additive reproduction will be described further in Chapter 7.

This process is shown schematically in Figure 14, which separates the intensity mapping from the RGB mapping. The intensity mapping is specified as the transformation from scene intensity values to displayed intensity values. The color component of the reproduction process transforms the red, green, and blue primaries used to encode and display the images. The result is not a precise copy of the original scene in any sense, but one that appears to be a good reproduction.

The output RGB color space for television is precisely defined with respect to CIE tristimulus values, typically specified as the chromaticity of each primary plus the white point. To minimize the transformations needed, television cameras are optimized for output to a television receiver.

The camera uses colored filters to convert the scene spectra to the output color space, as shown in Figure 14 (top). Ideally, these filters are color-matching functions based on the output primaries so the encoding will match human color vision, as described in Chapter 1. In this ideal case, the RGB values produced act as tristimulus values, giving a good visual match. Real filters are rarely ideal, however, so

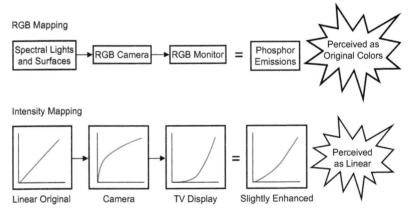

Figure 14.

The color reproduction process in television. Color is captured as RGB and reproduced as RGB. The intensity transfer function is captured linearly, encoded, and decoded slightly non-linearly to create the correct appearance.

errors in the color encoding are inevitable. Another source of error is colors that lie outside the gamut defined by the primaries, which is smaller and less saturated than the set of visible colors. Imagine trying to capture the colors of a laser show as an extreme example of this problem.

Figure 14 (bottom) describes how intensity is mapped throughout the color reproduction processes. The input pixels are encoded non-linearly with respect to intensity in a way that can be automatically decoded by the non-linear transfer function of the television display. This encoding takes the approximate form of an inverse gamma curve, to match the native gamma curve of the TV display. Note, however, that the desired result is not a linear function, but one that is slightly curved, which produces a picture with slightly enhanced contrast and saturation. This was designed to accommodate the difference in appearance between a television in a dimly lit room and a brightly light natural scene, as was recommended by Bartelson and Breneman (see Chapter 2).

Compression is very important in video in order to manage the enormous amount of data generated. As a first step, the RGB signals are transformed into a luminance channel and two color channels. This type of encoding creates a separate image that contains all of the edge information needed to reproduce details in the image, and also provides a grayscale image as one of the channels. The color channels can be more strongly compressed than the luminance channel without loss of visual information (for more detail, see Chapter 6).

Video encoding compresses the signal to the point that both color and shading artifacts are visible, especially if a single frame is viewed as a still image. Much criticism of video color comes from the compression, and also from incorrect encoding and decoding of the video signal.

Given that there is a formal specification for video color, why do all of the televisions in a store display the same program in different colors? While this is often used as an example of why color reproduction is hopelessly difficult, it actually demonstrates how much color reproduction principles depend on the application. People generally only watch one television at a time, and without any reference, any of the displayed images is a "good enough" reproduction. This may be less a matter of ignorance than of visual adaptation at work. The dim blue color of a television viewed through a window is an indication of its actual color properties, but the visual system will conveniently adapt to make it look as real as life when it is the primary source of visual information. In this way, perception reinforces the idea that big, bright, and cheap is much more important than a precise color match for consumers at large. If there were a need to control the color more tightly, it could be done; professional studio monitors are much more tightly controlled and accurate. All it takes is money and the discipline to maintain the standard settings.

In summary, a television has a much smaller gamut, both in color and brightness, than the original scenes it evokes. The signal is encoded and compressed to an amazing degree and then displayed on a consumer product sold by the millions. Staged television can be designed to compensate for this process, but news and other "live" footage must be continuously adjusted, often in real-time, to provide an acceptable reproduction. Think about this when you watch the evening news.

## Graphic Arts Example

In the print-oriented branch of the graphic arts, input is traditionally scanned photographs or slides, whose spectral representations are limited to the specific colorants used to create them. This makes them easier to capture accurately than the natural scenes captured by a video camera. Output is a four-color printing press, a subtractive color reproduction system that is much less predictable and controllable than an additive display. In between, digital prepress systems are used to crop, edit, and combine images, to add text and other line art, and to generally prepare the complete printed page or set of pages that will be printed together. Subtractive color reproduction, including printing, will be described more completely in Chapter 8. This overview is to show the color reproduction process.

Figure 15 shows a traditional prepress pipeline. At both ends of the process are images generated by mixtures of subtractive primary

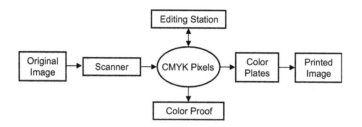

Figure 15.
Color reproduction pipeline in the graphic arts, including the option to edit the CMYK digital representation.

colors. Graphic arts scanners use red, green, and blue filters for image capture, but convert immediately to cyan, magenta, yellow, and black (CMYK) pixels, which are defined to match the characteristics of the printing press. Note that black is added selectively to make the dark colors darker, and does not define a grayscale representation of the image. These pixels are used to create a color proof, or as input to an editing station, and ultimately to make the color plates that apply the ink during printing.

The key component of graphic arts color reproduction is the transformation from the scanned input, which is an RGB process, to CMYK output. The CMYK representation is a process-specific one, where the pixel values represent the ink values at each point. Once this representation is produced, control of the color reproduction process reduces to ensuring that the specified values are accurately reproduced throughout the mechanics of the printing process. This process includes plate making, proofing, and the actual printing itself. This is conceptually simple, but can be technically challenging, given the mechanics of plates, ink, and paper.

The tone reproduction pipeline for printing is shown in Figure 16. The original linear reflectance or transmittance values are transformed in the "Tone Mapping" step to create linear CMYK pixel values. Often, the original values must be compressed, especially if the original is

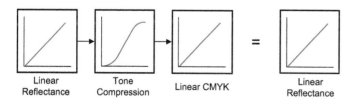

Figure 16.
Tone reproduction in printing. Linear reflectance values are mapped to linear CMYK pixels and printed.

a transparency (color slide). This is the aesthetic/perceptual part of the process, which is image-dependent. One common mapping is shown. This form will maintain linearity throughout most of the tone range, compressing only the very dark and very light colors. The linear CMYK pixels create linear reflectance values on the printed page.

The primary color transformation in the graphic arts pipeline is in the scanning step, where the original image is converted to printer CMYK. Rather than mimicking color matching functions, scanner filters are usually tuned to select the primary colorants used in the original photograph. If the filters were ideal, and the input CMY were identical to the output CMY, the transformation would be trivial. However, many factors add complexity to this process. The capture is rarely completely accurate. The CMY in film is not the same as the CMY in printing inks, creating a gamut mismatch. Subtractive primaries do not add ideally—there is cross-talk between the colors that makes it difficult to predict the color of their sums. Black must be added selectively to some colors to improve contrast.

In the traditional graphic arts, a skilled scanner operator controlled all of the color transformations, applying the basic principles introduced at the beginning of this chapter. Different processes and applications require different color transformations and different degrees of color fidelity. In the glory days of the Sears catalogue, for example, the color match between the printed colors and the original objects was superb. At every step in the process, craftsmen would make fine adjustments to match the color to the original objects, which were provided for reference. Today, measurement and automated control are becoming as important as such careful craftsmanship, although skilled expertise is still needed in many aspects of printing and publishing.

A color proof is a print made to mimic the final output. It is impractical in most cases to proof on the actual press, so various technologies have been developed for proofing. Color adjustments are often made based on the proof, until the customer signs off on the result. It is then the job of the press operators to adjust the printing press until the printed result matches the proof. For more critical applications, the customer may also do a "press check" to view the color as it is printed and to provide feedback as the press is adjusted for the final run.

The first graphic arts editing stations were expensive, custom systems used primarily for cropping and combining images. The RGB representation generated so that the images could be displayed on a monitor for editing only approximated the true colors, and was not used for color evaluation. While graphic arts professionals rely more and more on displays, a displayed image still cannot provide the color fidelity of a subtractive proof. Desktop programs such as Adobe Photoshop and Quark Express have replaced customized systems, and are also moving many of these functions away from the prepress house and into the hands of independent graphic artists.

In 1982, digital prepress systems were highly specialized and cost hundreds of thousands of dollars. Ten years later, cartridge disks containing images and layouts from desktop publishing programs had become a routine part of the prepress process, and what was once a closed, craft industry, was starting to be practiced by anyone with access to desktop publishing software. Today, an all-digital path from camera to press is not only possible, but is fast becoming the standard method for many forms of publication. The output is still paper, but it may be printed at Kinko's, or even on a desktop printer, rather than on a printing press. And, the output may not even be primarily paper—it could be the web, or some digital disk format such as CD-ROM or DVD. The star-shaped workflow shown in Figure 7 is a more accurate description of many printing and publishing environments than the process in Figure 15.

# Desktop Publishing Example

Desktop publishing is not a traditional craft industry like television or printing, but it is probably a form of color reproduction familiar to most readers of this book. This example will show how the principles introduced in this chapter can be applied to desktop publishing by creating a process centered around RGB pixels. This model will be described in a qualitative manner here—a more metric version can be achieved using a color management system, as described in Chapter 9.

In desktop publishing, input is usually created with a desktop scanner, a digital camera, or with an image editing program; output is usually a desktop printer or the web. Desktop scanners and printers come with default transformations to and from RGB pixels, which are viewed on a color display as part of the editing composition process. For the user, the "original" image is the one on the display, as it is often easier to adjust the displayed image than to adjust the input and output transformations. This suggests a display-centered view of the color reproduction process, where all transformations are defined relative to the user's display.

Let's look at what this means in more detail, using the process described in Figure 17. In this example, input from a scanner or a camera are combined with colors synthesized in an application, then directed to a specific printer for output. The goal, from the user's standpoint, is to make the printed colors match the display.

Figure 17.

Color reproduction for desktop publishing. This example is centered on an RGB representation, which is viewed on a digital display.

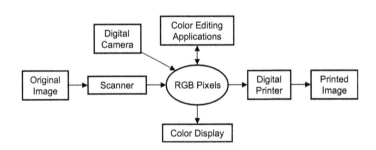

First, the input, output, and display components of the system must be stabilized so that their color characteristics are predictable. This involves establishing standard settings for all hardware and a standard set of drivers and their parameters. It also means controlling the lighting, which affects the appearance of colors. For prints, whose color depends critically on the lighting, there should also be a standard viewing environment with consistent lighting. The display should have little or no light shining directly on it, and be used in a room that is dim but not completely dark.

The transformations from input to display to output must also be stabilized. This involves choosing which applications will be used for input, viewing, and printing, as well as selecting any settings provided by these applications. A good way to do this is to choose some representative pictures, plus some simple test patterns of solid colors and steps of gray, to scan, display, and print. Since the scanned print, the display, and the output printer all have different gamuts, color shifts are inevitable. Adjusting whatever parameters are available until reasonable results can be achieved with these standard images defines the default tone reproduction and color balance characteristics for the reproduction process. The same test images can be used for process control; the system is in the correct state if processing them creates the same result.

Inevitably, the standard setup will not give optimal results for all images. The cleanest approach is to make any image-specific corrections by editing the image itself, using some form of image editing software. This doesn't change the default setup and creates an image file that should always print in exactly the same way.

The alternative to editing the image is to systematically adjust the input or output transformation on a per-image basis. This can be attractive if the printer or scanner has well-designed controls. On input, changing the settings can be considered a form of image editing; the results are encapsulated in the resulting file. Changing the output settings, however, is a separate process that must be performed each time the image is printed. Often, systems with complicated settings provide a way to save them to a file. This is always an excellent idea, both for consistency, and to document the settings.

In summary, the goal is to establish a consistent process, verify it with standard images and patterns, then hold this process as stable as possible. All aesthetic adjustments should be made to the image itself whenever possible. If color management is available, as it is in most modern desktop environments, turning it on should significantly improve the default color transformations. It is especially helpful for the transformation from display to print, which is the most complicated part of this example.

# Summary

As was stated in the introduction, good color reproduction is both an art and a science. There is rarely a single definition for a "good" reproduction, which is affected by factors both technical and economic.

The color reproduction principles presented in this chapter evolved from the aesthetic reproduction of images that evoke the known world. Primarily input as photographs, these images are the driving force in the traditional color reproduction industries. They have an impact on the design of cameras, film, televisions, and printing systems. Artistic, illustrative, or scientific images may not match these assumptions, and reproducing them with traditional processes may produce unexpected results. My favorite story illustrating this is from the early days of computer graphics. A group of researchers had worked hard to develop models of fog for a simulation of the New York harbor. When they sent pictures to a printer for reproduction, he helpfully adjusted the tone reproduction to improve the contrast, thereby removing all the fog.

As digital color media evolve, there is a move away from the traditional point-to-point color reproduction systems to a many-to-many mapping. While point-to-point is hard, many-to-many is much harder. Device-independent standards for color interchange are crucial in an all-digital world.

The conversion of the graphic arts from a chemical and film-based process to a digital one was fast and revolutionary. A similar revolution will eventually overtake all color reproduction industries once the basic technologies achieve a sufficient quality level and digital tools have sufficient power to manage standard volumes of data. Snapshot quality consumer-grade cameras are rapidly overtaking film-based ones, and high-end systems are making inroads in portrait studios. Digital special effects and post processing has made digital color a key part of movie-making, and theater-sized digital projectors offer an all-digital option for cinema. Digital video is replacing analog, especially for applications other than broadcast. These developments emphasize the need to create standards for digital color specification and encoding and tools for effectively managing color across media.

# 5  Color Reproduction

# 6
# Color Image Capture

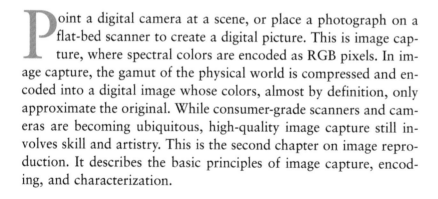

Point a digital camera at a scene, or place a photograph on a flat-bed scanner to create a digital picture. This is image capture, where spectral colors are encoded as RGB pixels. In image capture, the gamut of the physical world is compressed and encoded into a digital image whose colors, almost by definition, only approximate the original. While consumer-grade scanners and cameras are becoming ubiquitous, high-quality image capture still involves skill and artistry. This is the second chapter on image reproduction. It describes the basic principles of image capture, encoding, and characterization.

## Introduction

Image capture is the first stage of the color reproduction process. It is the stage where the lights and colors in the original scene are encoded as a rectangular array of numbers. These numbers represent the brightness and color at each point in a planar image, and are the

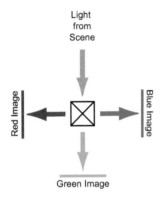

Light from Scene

Red Image

Blue Image

Green Image

Figure 1.

Colored light becomes red, green, and blue pixels in image capture.

familiar pixels used in imaging systems today. This is summarized in Figure 1. Colored lights of various spectral distributions become red, green, and blue pixels. Ideally, image capture would accurately capture the full visible gamut of colors. In practice, the captured colors are only an approximation of what the eye would see.

There are two basic forms of image capture technology: cameras and scanners. Cameras (photographic and video) are used to create images directly from lights and objects—they must create an image from a 3D scene, converting arbitrary spectra to colored pixels. Like the eye, they must adapt to a wide range of lighting conditions.

Image scanners, such as those used in the graphic arts, start with flat images that are most often photographs or prints. The input colors are usually composed from a limited set of dyes or inks, so the range of spectra is more limited, as is the input color gamut. As a result, these systems can be tuned for specific input spectra, simplifying image reproduction and characterization.

A captured image must be encoded and stored in a digital imaging system. The raw data captured by the hardware are processed and often compressed, especially for camera forms of capture. The encoding specifies the color and position for each pixel, ideally in a way that will enable it to be accurately and aesthetically reproduced. While older image formats have no colorimetric specification, more modern ones use the principles of device-independent color used in color management systems.

Image capture is described in any book on color image reproduction. Hunt's book, the *The Reproduction of Colour,* has technical detail on both cameras and scanners, and Poynton's *Digital Video* discusses digital video capture in full technical detail. Green and MacDonald's *Colour Engineering* includes a chapter on camera capture for device-independent color.

This chapter starts with a discussion of the properties of RGB filters for image capture, followed by an overview of image capture technology such as digital cameras and image scanners. The following sections discuss image encoding, concluding with a description of characterization methods for image capture systems.

116

# Fundamentals of Image Capture: RGB Filters

To perform image capture, a scene is viewed through filters that separate the image into its component colors, red, green, and blue, which are then rasterized into pixels to create a colored image. To give these colors meaning, we must either relate them directly to perceptual quantities in the visual system or match them to a particular output system such that the output image matches the input image. Traditionally, capture was tightly tied to a specific output medium.

In Chapter 1, we discussed how human vision encodes colors into three values. An image capture system designed on these principles would use the same sort of color-matching functions used in color measurement; that is, the capture filters would be color-matching functions, and the resulting red, green, and blue values would be tristimulus values. Such filters are called *colorimetric filters*.

Filters used in television are not general color-matching functions, but act as color-matching functions within the gamut of the television display. They miscode colors outside of the gamut, but as these cannot be accurately displayed anyway, it matters less.

Film, in contrast, consists of dye layers that are selectively sensitive to non-overlapping red, green, and blue regions of the spectrum. These "filters" are optimized for the colors produced when the film is developed. Their design is based on the principles of subtractive color reproduction rather than the principles of color matching.

Filters used in image scanners for the graphic arts are often narrow band filters tuned to capture the major colorants used in photographic transparencies or prints. These are not at all like general color-matching functions, but are tuned instead to encode the colors as three independent separations in a way that optimizes their reproduction via printing. In this way, they are like the filters used in color densitometry.

Digital Still Cameras (DSC) originated from video systems. However, their evolution into consumer products has placed different demands on their filters which must be low-cost and optimized to capture the maximum light on a small area. In addition, many DSC images are ultimately printed, rather than viewed on a specific display. As a result, few DSC systems use colorimetric filters, or even the gamut-limited colorimetric filters used in video.

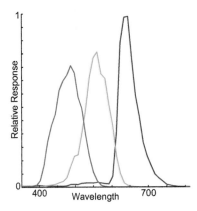

Figure 2.

The spectral sensitivity curves for a Sony imaging array after filtering out unwanted infrared wavelengths. (Data courtesy of Paul Hubel.)

Figure 2 shows a set of spectral sensitivity functions for a Sony imaging array, as would be used in a digital camera. Comparing these curves to the CIE color-matching functions described in Chapter 1 takes more than a simple visual inspection. Remember, there isn't a single set of color-matching functions, but a family of functions that are all linearly related; the CIE color-matching functions are but one example. Fully evaluating how well a particular filter set approximates a visually accurate color-matching function must include finding the linear transformation that gives the closest match, then making the comparison.

Many image capture systems take advantage of the strong correlation between a green filter and the luminous efficiency function of the eye, which is shown in Figure 3. As discussed in Chapter 2, the area under the luminous efficiency function is the luminance of the color, which quantifies its apparent brightness. The eye is more sensitive to wavelengths in the green-yellow part of the spectrum, so a "luminance filter" will appear green. As a result, the green signal contains the most perceptually important brightness information.

Figure 3.

The luminous efficiency function, whose area equals luminance, compared to a green camera filter.

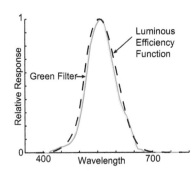

Most filters are dyes that selectively absorb or transmit different wavelengths, as described in Chapter 4. *Dichroic filters* or *dichroic mirrors* selectively transmit or reflect different wavelengths of light, effectively splitting the light into different colored components. These components can then be directed at subsequent dichroic filters for further filtering. In this way, no light is lost to absorption.

Color filters that are not colorimetric will miscode some colors relative to human vision. That is, two colors that look different to a human viewer may encode to the same RGB triple, and vice versa. These differences can be quite dramatic, such as a blue flower that photographs pink because it reflects long-wavelength red light that is visible to film but not to the eye.

A digital example can be seen in Figure 4, which shows a picture of a plasma display and a set of three laptops with liquid crystal displays. All appeared a similar deep, royal purple when viewed directly, but the digital camera used to take this digital photograph has

Figure 4.

An example of color miscoding by non-colorimetric capture filter. The large plasma display and three liquid crystal laptop displays were visually all the same color, but the camera encoded them very differently.

encoded the spectrum from the plasma panel in a way that it appears bright blue, whereas the laptop displays appear a dark, reddish brown. These different spectra, which appear to be the same color (*metameric matches*), encode to different colors when viewed through a digital camera because the camera filters are not color-matching functions.

## Camera Systems

Digital camera systems contain optics that image light on a sensor, typically a CCD array with filters, though some newer models use CMOS arrays. The resulting image is encoded and usually compressed, either in video format or as a JPEG image. The quality of the resulting image depends on the characteristics of the filters, the imaging electronics, and the associated processing.

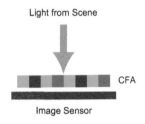

Light from Scene

CFA

Image Sensor

Figure 5.

A schematic of image capture in a single-element digital camera, with a CFA over the imaging sensor.

Figure 6.

The layout of the Bayer color filter array, one of the most common CFA arrangements.

The most accurate color capture electronics use a dichroic filter to split the light into its red, green, and blue components. Each component is imaged separately onto an array of sensors, then combined to create a full color image, similar to Figure 1.

Most digital cameras, however, use a single imaging element and a tiled Color Filter Array (CFA), which is more compact and less expensive than a three-element system. This is illustrated in Figure 5. Each pixel is initially either red, green, or blue; adjacent pixels must be interpolated to create a full color value for each pixel, a process called *image reconstruction*.

A common CFA, called the Bayer array, is shown in Figure 6. In the Bayer arrangement, 50% of the filters are green to maximize the capture of luminance information, which increases sharpness.

For small, snapshot cameras, the combination of the CFA and the compression applied to the image can create unsightly artifacts in the image, especially in the dark regions. Larger, more expensive cameras provide the option to bypass the compression step, but may still show artifacts due to the CFA, especially at sharp edges.

A recent innovation from Foveon uses a layered, rather than a tiled, approach to creating a camera with a single imaging element. Light is absorbed at different depths in the CMOS array, as shown in Figure 7. This is quite similar to the way photographic film works, with its layers of dyes, as described in Chapter 8.

The resolution of the imaging element defines the resolution for the resulting image. Obviously, the higher the resolution, the more detail possible in the image. Raw resolution numbers, however, do not tell the entire story. The camera optics, the sensitivity of the detectors, the characteristics of the filters, and sophistication of the image reconstruction algorithms all have a strong impact on the resulting image quality.

## Exposure and White Balance

The human visual system adapts to variations in brightness and white balance (see Chapter 2). In a camera, the exposure is the brightness adaptation mechanism, and white balance is the chromatic adaptation mechanism. The quality of these adaptation mechanisms are critical to the quality of the captured image.

In traditional photography, exposure and white balance are controlled by a combination of film choice, shutter speed, lens filters, and aperture size, plus the addition of extra lighting, such as the flash. A professional photographer measures the light to determine its brightness and color temperature, then select film, exposure, and filtering to get the best results. Further adjustment can occur in the darkroom. For a snapshot camera, the choices are usually reduced to the choice of film and the automatic system that determines whether to trigger the flash.

In digital photography, there is no choice of film. The same sensors and capture filters must be made to accurately record all scenes. Like film cameras, shutter speed and aperture size control exposure in digital cameras. Similarly, the flash can be automatically triggered or manually applied. White balancing is achieved algorithmically.

Automatic White Balance (AWB) algorithms first identify a white or neutral color, then adjust the power of the red, green, and blue components so that equal values are neutral. The simplest AWB algorithm averages all the pixels in a scene and assumes the resulting color is gray; clearly, this approach will fail for many pictures. More complex, but more effective algorithms are often proprietary. Better AWB algorithms are the subject of on-going research.

An effective mechanical solution for white balancing is to have the user point the camera at a known white object during a setup step—some cameras include a white target on the lens cap that can be included in the picture for this purpose. Simpler, but less general, is to have the user select from one of several standard light sources, usually incandescent, fluorescent, or daylight. One of the primary differences between professional and consumer digital video and digital cameras is the degree of control provided for white balance, exposure, and adjustments that improve image quality.

Digital images can easily be post-processed on a computer (the "digital darkroom"). Given the complexity of exposure and white balancing algorithms, the simplest camera design would leave ev-

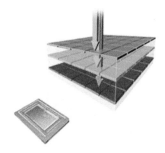

Figure 7.

The Foveon layered sensor, which doesn't use colored filters. Different wavelengths travel to different depths in the silicon material, where they can be read out to construct the red, green, and blue images. (Image courtesy of Foveon.)

erything to post-processing; however, the image would need to include sufficient color and brightness information so that it could be adjusted without introducing artifacts, and any image displayed on the camera's previewer would not show the results of the compensation. A system embedded in the camera can perform compensation as the picture is taken, adjusting exposure time, for example, so that all the grayscale levels can be encoded in eight bits, and the final image can be JPEG compressed and stored compactly. But, more local processing means a more expensive camera with more power requirements (which shortens battery life) and more time needed to process each picture. This tradeoff is a significant factor in the engineering of digital camera systems.

# Image Scanners

Image scanners consist of a drum or platen that holds the picture to be captured, a light source, and a sensor containing colored filters. The sensor travels across the picture, capturing pixels in sequence to create an image, as illustrated in Figure 8.

Most scanners are designed with a sensor array for speed. A desktop scanner typically has a bar of sensor elements the maximum width of the image. The colored filters are arranged in three rows, one for each of red, green, and blue, as illustrated in Figure 9. The resolution of the sensors on the scan bar defines the resolution of the resulting image in one direction, the motion of the bar defines the other.

Image scanners are products of the graphic arts industry, where their primary application is scanning photographic media for printing. As a result, the red, green, and blue filters in professional scanners are designed to selectively measure

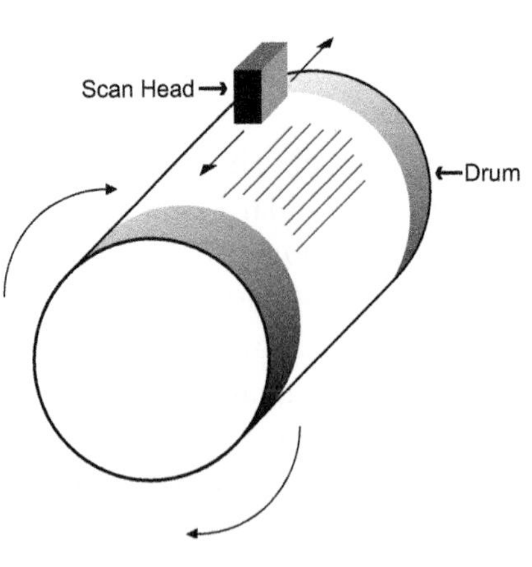

Figure 8.

On a drum scanner, the drum spins while the sensor head scans across to create the image.

Scan Head →

← Drum

the three colorants in the film stock. Different filters are used for prints, positive transparencies, and negative film. The trend in desktop scanners, however, is to use more general filters like those in digital cameras.

Images from scanning systems are usually encoded as RGB reflectance or transmittance at each pixel. To capture the full dynamic range of film requires at least 10 bits/pixel; some systems produce 12 or even 16 bits/pixel. Camera systems, in comparison, generally create a non-linear encoding of brightness, which requires fewer bits/pixel.

The quoted spatial resolution on image scanners is very high, often over 1000 pixels/inch. This serves two purposes: it allows the scanner to be used for capturing the sharp edges of lines and text, and it allows a small image to be scanned and scaled without loss of sharpness. The ideal resolution depends on the application. For displays, there is a one-to-one correspondence between the scanned pixels and the displayed pixels. Typical good-quality computer displays run around 100 pixels/inch, though 72 pixels/inch is still often quoted for "display resolution."

In printing, the image pixels are sampled and printed in a high-resolution pattern called a *halftone pattern* to create the perception of grayscale. Because of this resampling, there is no need to precisely match the scanned resolution to the printer—there simply needs to be sufficient information to reproduce the detail in the image. The halftone screen frequency is not the printer's native resolution, but is defined by the algorithms that create grayscale from the binary colors on the printer, as described in Chapter 8.

For printing continuous tone color images, the rule of thumb is that the original image should be roughly twice the resolution of the halftone frequency, though as low as 1.2 times the halftone frequency is often sufficient. Halftone screens for offset printing are typically 150–200 lines/inch, which means no more than 300–400 pixels/inch (at the final printed size) is needed for continuous tone imagery. Desktop printers don't use halftone screens in the same way, and their resolution is lower, so even less image resolution is needed for these systems. The way to evaluate the resolution needed for a particular printer is to scan an image with

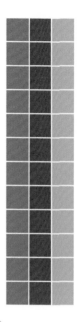

Figure 9.
A typical filter array for a scan bar, as would be used in a desktop scanner. The bar of sensors defines a full row of pixels; the bar scans across the image.

fine detail, such as hair or fur, then print successively lower resolution versions of the image until you see the detail degrade.

High-contrast images containing text and lines, however, must be scanned at very high resolution to preserve sharpness. In this case, the scanning resolution should match or exceed that of the printer. In general, it is best to use the highest scanning resolution available for this type of imagery.

The quality of an image scanner depends on much more than its resolution and bit depth. Noise, non-uniformity across the sensor bar, and other optical distortions can degrade the image quality, even at high resolutions.

# Image Encoding

In the digital domain, a captured image is saved as a file in an image format such at TIFF, PNG, or JPEG. These represent a two-dimensional array of pixels, where each pixel encodes the red, green, and blue component of the image at that point. Pixels are usually encoded as 8–16 bit values, with 8 bits by far the most common. What these numbers mean with respect to any physical or psychophysical specification of color, however, is too often left unspecified in digital image formats and applications.

To fully specify an image pixel, we need a precise specification for the primary colors, and a brightness function that maps pixel values to some measurable quantity such as intensity or density. An RGB representation can be precisely specified by defining the primaries in terms of their CIE tristimulus values and the functions that map from pixel values to luminance or intensity, as described in Chapter 3.

A CMYK image, in contrast, specifies percentages of ink values for printing. To be precise, it must include specifically which inks are to be used; there is variation in the color of "cyan" just as there is in the color of "red." However, there is no simple transformation to a device-independent specification based only on the ink colors, as will be described further in Chapter 8, Subtractive Color.

# Difference Encodings

Many image formats take advantage of basic perceptual principles by using an encoding that combines a brightness value with two color values. These are usually simple transformations of the RGB signals, with the brightness component proportional to luminance (Y). The primary advantage of this form of encoding is that the brightness channel contains all the critical edge information that defines the sharpness of the image. It is possible, therefore, to reduce the spatial encoding of the chromatic channels by a factor of 2 or more, thereby reducing the image size without visible loss of quality. Furthermore, the brightness channel automatically provides a grayscale representation of the image. Figure 10 shows a full-color image separated into an achromatic and two color-difference images, in this case, CIELAB. Even with the transformation applied by the printing process, the L* image looks like a grayscale version

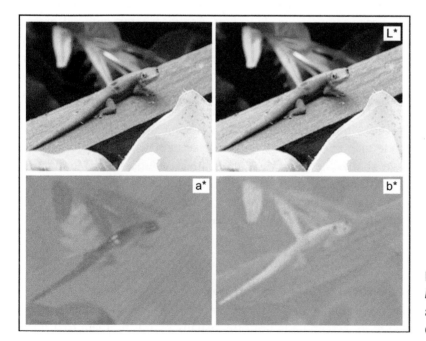

Figure 10.

A color image separated into its achromatic (L*) and color difference (a*, b*) components.

Figure 11.

The process of converting an RGB signal to a color difference encoding.

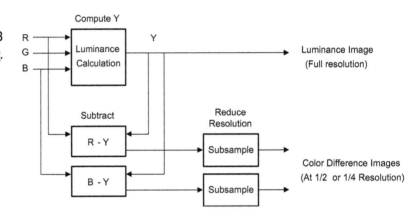

of the color images. The color-difference images (a* and b*) are almost impossible to identify.

Let's look a little more at this type of encoding, which was first extensively used in television and video signals, and is also used in most image compression algorithms. Assume we start with red, green, and blue intensity values of a known chromaticity that are linear in intensity. Luminance can be extracted from these values using a linear transformation based on the color of the RGB primaries. Two color-difference images are then created, usually Y-R and Y-B. These images can be resampled to a lower spatial resolution and encoded with the luminance information; this process is summarized in Figure 11. To decode such an image, we must first interpolate the color difference images so that they have the same spatial resolution as the luminance image. Then, for each pixel, R, G, and B can be recovered by inverting the transformation above; first, subtracting the luminance to get R and B; then, inverting the luminance transformation to recover G.

# Non-Linear Encodings

To minimize the quantization that would occur by encoding them as 8-bit pixels, linear intensity values are encoded non-linearly to make them more perceptually uniform, usually by raising them to a frac-

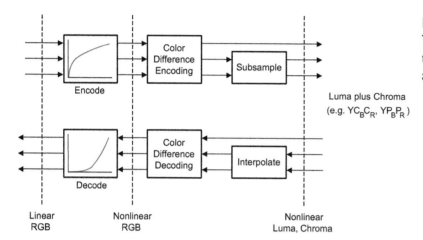

Figure 12.

The process of converting linear RGB to a non-linear difference encoding, as in video encodings.

tional power. This is similar to the encoding in perceptual color spaces that applies a similar function to create the perceptually-uniform L* metric. Most effective image encodings include some sort of non-linear encoding, in contrast to the linear encodings common in computer graphics.

The example in Figure 11 used linear RGB values. Figure 12 shows the addition of a non-linear encoding/decoding step, as in a video color reproduction system. The input values are first raised to a power of roughly 1/2.2 (depending on the standard used). Then, they are transformed into a non-linear luminance value (called *luma*) and two non-linear color-difference values (called *chroma*). The chroma images can be subsampled as above. To invert this process, the chroma images are first interpolated and then converted, along with the luma image, back to non-linear R, G, and B signals, which are passed through the inverse of the non-linear encoding function to create linear RGB values. In television and video, the non-linear decoding function is supplied automatically by the monitor response, as described in Chapter 3.

It is important to understand that the brightness channel, Y, in a non-linear difference encoding is *not luminance*, even though the notation is the same. It is luma, which is computed from non-linear R, G, and B values. Some references, especially in computer graphics, suggest using the luma equations from video encodings to compute "luminance." While this often generates a plausible grayscale image, true luminance can only be computed from linear values of RGB of known chromaticity. This will be discussed in more detail in Chapter 10.

# Encoding Formats and Standards

Well-specified file formats are an important aspect of image encoding. There are many different formats supporting a range of bit depths, compression methods, and more recently, formal color specifications. Those mentioned here are the most commonly used in color reproduction applications.

TIFF (Tag-based Image File Format) was first published by Aldus Corporation in 1986 after a series of meetings with a number of scanner and software developers. Adobe Systems, who eventually acquired Aldus, maintains the current specification, version 6.0. Its primary goal is to provide "a rich environment within which applications can exchange image data." It is structured with a small number of required tags, a larger number of standard extensions, and the option to register and include custom tags. One of the standard extensions is RGB image colorimetry. TIFF is a non-proprietary image format widely used in the graphic arts industry. It is flexible enough to encode bitmaps, grayscale, indexed color, RGB, CMYK, and specialized encodings such as high-dynamic range RGB.

The Adobe Photoshop file format is a widely used proprietary format that by default include precise color specification for both RGB and CMYK images, as well as grayscale, duotone, and other specialized encodings.

PNG (Portable Network Graphics) was designed as a patent-free replacement for CompuServe's GIF (Graphics Interchange Format) for images on the web. It is now a recognized format in many applications, and includes gamma and colorimetric information as required parts of the specification. PNG is recommended by the World Wide Web Consortium(W3C) as an image format. As well as being a web format, it can replace TIFF in some image editing applications.

Unlike TIFF, there is intended to be little variation in PNG encodings, making it easier to guarantee consistent application and display of PNG images.

While both TIFF and PNG support compression, JPEG is the most common compressed image format, especially now that it is the usual file format for digital camera output. JPEG stands for the Joint Photographic Experts Group, an interdisciplinary group of professionals working on imaging standards. The ubiquitous JPEG file combines the ISO 10918-1 (ITU-T T.81) standard for still image compression with a file format (JFIF) placed in the public domain by C-Cube Microsystems. JPEG is supported by most web browsers, imaging applications, and digital camera systems. There is no chromaticity information in a JPEG file, and while the pixels defined in the JFIF format are supposed to be linear with respect to intensity (as in Figure 11), it is reported that, in practice, they are usually encoded non-linearly (as in Figure 12), and the precise function used is dependent on the implementation. JPEG2000, the latest JPEG imaging standard, includes a complete specification for RGB color similar to that used in color management systems. Unfortunately, like TIFF, this information is included in the optional extensions part of the standard.

The EXIF (Exchangeable Image File) format is a JEIDA (Japan Electronic Industry Development Association) standard used in digital photography. The latest version combines JPEG compression and file format with the sRGB colorimetric specification (described in Chapter 3).

Photo CD was developed by Eastman Kodak company "for the platform-independent storage and retrieval of images captured by film and digitized by a film scanner." Photo CD discs are usually created by sending a role of film to a special Kodak site for scanning and encoding. The Photo CD format includes multiple resolutions, and a precise colorimetric specification, PhotoYCC.

Flashpix is an I3A (International Imaging Industry Associates) initiative to create a flexible, robust image format for a wide range of imaging applications. Like Photo CD, it includes several image resolutions and supports colorimetric specification in various formats, including PhotoYCC. There is an open source toolkit for manipulating Flashpix images available from the I3A website.

MPEG (Moving Picture Experts Group) was established around 1990 to create a standard for delivery of audio and video. There are three MPEG standards: MPEG-1, MPEG-2, and MPEG-4. The video portion uses the $YC_rC_b$ luma/chroma color space defined by the International Telecommunication Union for component digital video (ITU-R BT.601).

## Characterizing Images and Image Capture Systems

Image capture systems do not have gamuts in the sense that output systems do. Individual images have gamuts defined relative to the image encoding. The problem, then, is mapping from the original colors to the encoding to create a device-independent representation for the image. This would be easy for an image capture system that contained color-matching functions as filters; as these are rare, approximations must often be used. Clearly, the more that is known about the input, the more accurate the approximations can be.

Most image scanners are characterized with a chart such as the IT8.7/2 photographic chart shown in Figure 13. Each patch has a

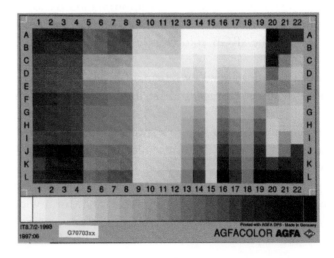

Figure 13.

An IT8.7/2 test chart used for characterizing a scanner. (Image courtesy of Agfa-Gevaert.)

known CIELAB value. Scanning the chart creates a correspondence between the resulting RGB pixels and the CIELAB values. Other values can be interpolated from this set, and the inverse transformation generated by inverting the table. IT8 scanner charts, or *scanner calibration charts*, are readily available in a variety of forms, primarily photographic prints and slides.

A characterization created using an IT8 chart, or any similar chart, is accurate only for the spectra set represented by the test chart because image scanners do not use colorimetric filters for their scanning. A characterization for photographic prints, for example, would not be as accurate for paint, or even for photographic slides. To be most accurate, the chart needs to be created from the type of materials being scanned. Figure 14 shows a hand-painted calibration chart created by a background painter at Warner Brothers. This chart, which reflects the color mixtures (and therefore the typical spectra) used by that particular painter, is used to characterize the scanner before that artist's paintings are scanned. Using a customized chart gives a much more accurate characterization than using a general-purpose one.

Characterizing cameras is harder than characterizing scanners because there is no constraint on the input spectra viewed by the

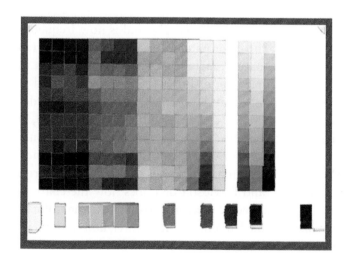

Figure 14.
A hand-painted IT8 test chart used to calibrate the scanning of hand-painted background images for cartoon animation. (Courtesy of Warner Brothers.)

Figure 15.
A picture of a Macbeth ColorChecker
(plus a curious cat).

Figure 15.
A picture of a Macbeth ColorChecker
(plus a curious cat).

camera, and because the lighting is less controlled. A digital camera could be characterized by photographing an IT8 test chart, but the characterization would only be truly accurate for the material of the chart viewed under that particular set of lights.

The Macbeth ColorChecker, shown in Figure 15, was developed as a spectral reference for photography and is now frequently used in digital imaging. The color patches are paint whose spectra mimic common spectra in nature. Included are flesh tones; colors for foliage and flowers; a selection of bright, primary colors; and a set of uniformly spaced, flat, neutral colors. The ColorChecker defines a standard input source that can be used for process control, and for comparing capture systems. ColorChecker charts can be purchased at photography stores for around $80, or online from GretagMacbeth, who manufactures them. A 1976 article by McCamy, Marcus, and Davidson describes how the chart was designed, including the specific spectral distributions. The ColorChecker data is available online from the manufacturer, and is found in many color data resources.

Reference.
C.S. McCamy, H. Marcus, and J.G. Davidson. "A Color-Rendition Chart." *Journal of Applied Photographic Engineering* 2:3 (1976), 95–99.

To avoid the problems of characterizing image capture systems described above, it is common instead to characterize the image with respect to some standard display. This creates a characterization defined by the displayed colors, rather than the original spectral representations. This is exactly what video standards do—the filters are defined with respect to a fully-specified, standard display. The better the quality of the video camera, the better the colors will correspond to the display specification. Similarly, a digital camera user can only evaluate an image by displaying or printing it, and often modifies the image based on its displayed appearance. Therefore, it often makes good sense to treat the displayed image as the "original" and base its characterization entirely on the display properties.

The technique of using the display to characterize the image can also be applied to images captured by scanners. Many workflows in the graphic arts, however, primarily scan film and produce print. These materials have the property that some of their gamut is outside the monitor gamut. Characterizing the scanner with respect to the input print material by using a test chart makes it possible to capture and preserve these colors in the workflow.

## Summary

Image capture is the process of encoding spectral input as RGB pixels. Even ten years ago, image capture systems were expensive tools used only by professionals. Now, digital cameras and image scanners are consumer items. The evolution of digital cameras is particularly fast right now. They are making inroads on the snapshot camera market, as well as some parts of the professional market, such as portrait photography. As a result, there is a rapidly growing interest in digital color reproduction by people who only want to print their digital photographs. Many of these users are skilled neither in printing nor photography, yet would like to achieve at least "snapshot" quality prints of their digital photos. Alternatively, many are "prosumers"; primarily, skilled photographers entering the digital color market with more knowledge and higher standards for their prints. All of this is creating new pressure on color management systems, which have been designed and developed primarily for the graphic arts market. This market traditionally involves scanners, not cameras, and professionals whose color expertise is printing, not photography.

An electromechanical image capture system can only approximate the sensitivity and adaptability of our eyes. Not that the human eye is such a wonderful optical and imaging system in the sense of precision optics and high-resolution sensors. But, its ability to adapt and interpolate is unequaled. A mechanical image capture system must model some of this adaptability to control exposure and to correctly white balance so that the image has a good dynamic range and doesn't include a strong color cast. Since no input capture system is a perfect match to human vision, any attempt to create a correspondence between input spectra and perceptual quantities like CIE tristimulus values will depend on the specific properties of the calibration targets and the camera settings.

A typical image encoding format is very precise about the spatial arrangement of its pixels, but often mute about their photometric and colorimetric properties. The original JPEG specification, for example, includes no photometric or colorimetric information. To fully specify an image, there must be a specification in some absolute sense of the color of its primaries and how the digital value of its pixels map to some photometric or visual metric like intensity or luminance. Without such a specification, there is no way to create a consistent rendering of the image.

# 6 Color Image Capture

# 7

# Additive Color Systems

Additive color systems paint with light. Such systems include monitors, flat-panel displays (LCD, LED, plasma), and digital projectors. The traditional additive color reproduction industry is television, though graphics displayed on a monitor may be more familiar to many readers of this book. This is the third of five chapters on color reproduction. It describes the basic principles common to additive color systems, some technology-specific details, and some practical information about characterizing them to a device-independent color standard.

## Introduction

Additive color systems became part of the common experience with the advent of color television. Tiny colored spots of light, as shown in Figure 1, combine to create moving images that invoke some form of "reality." Alternatively, the colored light can be overlaid, as

Figure 1.

Tiny red, green, and blue spots of light tile a television display.

137

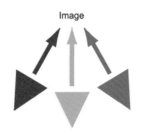

Image

Figure 2.

Additive color consists of combining red, green, and blue images.

indicated in Figure 2, which illustrates how a CRT-based projector creates its image. Digital projectors combine the image internally using mirrors, and the smallest digital projectors have only one imaging element and a color wheel. In this last case, the three colored images are added temporally. Whether presented as a mosaic, overlaid, or presented sequentially, it is adding together three glowing, red, green, and blue images that defines additive color.

Television and video are the additive color reproduction industries, which were originally designed as analog technology. While digital broadcast television remains mired in politics and problems unique to retrofitting a decades-old mass media infrastructure, recorded media is rapidly becoming all digital. Both digital video and DVD are irresistibly more convenient and provide much better image quality than their analog counterparts.

Digital color displays (whether desktop, laptop, or embedded) and digital projectors are the technologies that create additive digital color. Even five years ago, most digital color displays used Cathode Ray Tubes (CRTs), the same technology ubiquitous in television. More recently, flat-panel technologies are challenging the CRT, even in the television market; they are primarily liquid crystal displays augmented at the high end by plasma displays and modular rear-projection displays. Future technologies seek to make additive color displays as cheap as paper, big enough to cover a wall, and as bright as reality.

Lindsay MacDonald and Anthony Lowe have edited a good general reference called *Display Systems* that covers a number of display technologies. Charles Poynton's *A Technical Introduction to Digital Video* includes detailed information about video displays, and about the additive color reproduction process used in television and video. Charles has a second book, *Digital Video and HDTV Algorithms and Interfaces,* with additional and updated material. Stupp and Brennesholtz's book on *Projection Displays* describes various projector technologies. The Society for Information Display's (SID) conferences and publications are the standard venue for research in display technology.

Additive color is so fundamental to digital color that its principles were presented earlier in Chapter 3. The goal of this chapter is to tie those principles to technology and to color reproduction systems. After describing characteristics common to all technologies, the chapter provides some detail about different display and projector technologies. It concludes with a discussion of practical issues affecting the colorimetric measurement and characterization of additive color systems to support their use in device-independent color reproduction.

# Properties of Additive Display Systems

Additive display systems emit light as an array of colored spots that vary in intensity. Each displayed pixel is a combination of red, green, and blue light. The color gamut of an additive color system is defined by the color and brightness of its primary colors. The brighter and more saturated the primaries, the bigger and brighter the displayed gamut of colors. The three primaries sum to white, as shown in Figure 3.

The specific color of white depends on the color and relative power of the primaries: Making the blue stronger makes the white become more bluish, and vice versa. White points are often specified as color temperatures (5000, 6500, 9800), or as daylight equivalents (D50, D65, D98). The higher the temperature, the more bluish the color, as discussed in Chapter 4.

Commercially, display systems are described in terms of their size, spatial resolution, brightness, contrast, and the number of intensity levels produced at each pixel. The color gamut, if provided, is described by the chromaticity coordinates of the three primary colors plus the chromaticity of white and the display's maximum brightness.

Additive color gamuts are often illustrated by plotting them on a chromaticity diagram, where the primary colors define a triangle that encloses the display's gamut. The sRGB gamut, which is typical for a desktop CRT display, is shown in Figure 4. The largest additive gamut created from physical lights would use monochromatic laser colors as primaries; the corners of the triangle would lie right on the horseshoe-shaped spectrum locus. However, even this gamut will not encompass the entire visible spectrum; no triangle whose corners are on the spectrum locus can cover it entirely.

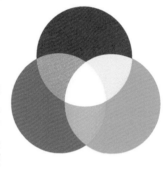

Figure 3.

The primary colors of an additive color system and their combinations.

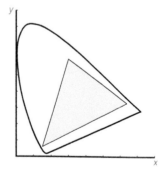

Figure 4.

Additive gamuts are triangles on the chromaticity diagram, as shown here for a typical CRT display (sRGB).

Using the terminology of Chapter 3, there is a brightness scale that describes the mapping from digital inputs to visible, measurable light for each primary color. When describing display systems, this brightness scale is usually characterized as a single *transfer function* that maps input values to measured intensity in steps from black to white. For optimal image quality, this should be a smooth function with no flat spots that runs from the darkest black to the brightest white for the display. Menu or button controls are usually provided to vary the transfer function in the display hardware. It can also be changed in the driving software.

Because additive systems emit light, they look best without any additional light shining on them. Ambient light, which is usually white light from room lights and windows, reduces both saturation and contrast, hence the tendency to turn down the lights when viewing projections and displays. The dimmer the display, the stronger the impact of ambient light.

The next sections discuss several features common to all additive display systems and their use in color reproduction systems, followed by a discussion of several different additive color technologies, including to what degree they match the ideal RGB model. The final section presents some practical advice on characterizing displays and projection systems to a device-independent color specification.

# Brightness and Contrast

The *brightness* of a display describes its light output. The metric for light power is the *lumen*, which is derived from the radiant power modulated by the wavelength sensitivity of the eye, the luminous efficiency function, as described in Chapter 2. Commercially, display brightness is specified in ANSI lumens, which is the average of nine measurements taken uniformly over the surface of the display. This averaging is necessary because displays vary in brightness over their surface, being typically brighter in the center and dimmer in the corners.

*Contrast* can be defined several ways, but for displays, it is usually provided as a ratio of the white-to-black brightness, such as 300:1. Contrast is sometimes used interchangeably with *dynamic range*, a term more common in signal processing. While CRT displays and projectors have very low brightness for black (effectively zero), flat-panel displays and digital projectors often have measurably high light output at black. As a result, while these technologies tend to be brighter than CRT-based ones, they do not necessarily have higher contrast, though the trend is in that direction.

Many display systems have controls marked "brightness" and "contrast" that change these properties by varying the transfer function. Brightness generally raises the entire function, whereas contrast stretches only the maximum brightness. These controls change the shape of the transfer curve, as shown in Figure 5. The black, dashed curve is a well-placed transfer function, a gamma curve similar to the transfer function of CRT displays. The red curve has its brightness set too low, such that dark values up to 0.4 all display as black. The blue curve has its brightness set too high (black not equal to zero). The yellow curve has both its brightness and its contrast set too high. Black is not at zero, and all colors above 0.8 are equal to white.

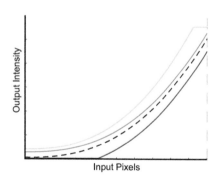

Output Intensity

Input Pixels

Figure 5.
An intensity transfer function maps pixels to measured intensity. The black, dashed curve is what is desired; it starts at black and smoothly progresses to white. The other curves will distort the image appearance, as described in the text and demonstrated in Figure 6.

Changing the transfer function can have a dramatic impact on the appearance of the displayed colors, as demonstrated in Figure 6. Assuming that Figure 6(a) is a good reproduction, corresponding to the black curve in Figure 5, Figure 6(b) is too dark and saturated, corresponding to the red curve. Figure 6(c) has its brightness set too high, so it lacks dark blacks and therefore contrast (corresponding to the blue curve). Figure 6(d) is too light, with burned-out highlights, corresponding to the yellow curve.

(a)

(b)

(c)

(d)

Figure 6.
The effect of the transfer functions in Figure 5 on an image appearance. (a) Good reproduction; (b) too dark (red curve); (c) black too high (blue curve); (d) burned out highlights, overall too light (yellow curve).

# Size and Spatial Resolution

Spatial resolution can be expressed either as pixels per inch or as total pixel dimensions, depending on the context. Given a physical size for the image, one can easily be converted to the other. Spatial resolution is defined by the combination of the display hardware and its digital drivers—most modern displays can operate at a range of resolutions, defined by the display controller in the driving computer.

For displays, spatial resolution is usually described by the total pixel dimensions plus the physical size of the display. For projection systems, there is a resolution defined by the dimensions of the imaging element, but the size varies with the projection. Digital display pixels are usually square (or close enough to approximate them as such), so that the horizontal and vertical pixels/inch are the same.

Some standard display resolutions are summarized in Table 1, along with their industry names. Most are in a width to height ratio of 4:3, though SXGA is 5:4. Most display systems will adapt to a range of resolutions, though the quality may suffer if the requested resolution is different than the native resolution of the hardware. Displays directed at the home entertainment market will support High-Definition Television (HDTV) formats, which are designed for

| Acronym | Width | Height | Ratio | Description |
|---------|-------|--------|-------|-------------|
| VGA | 640 | 480 | 4:3 | Video Graphics Array |
| SVGA | 800 | 600 | 4:3 | Super VGA |
| XGA | 1024 | 768 | 4:3 | eXtended Graphics Array |
| SXGA | 1280 | 1024 | 5:4 | Super XGA |
| UXGA | 1600 | 1200 | 4:3 | Ultra XGA |
| HDTV | 1920 | 1080 | 16:9 | High-Definition TV |
| HDTV | 1280 | 720 | 16:9 | High-Definition TV |

Table 1.

Standard display resolutions and names.

a ratio of 16:9 to make them compatible with movie formats. Only a few more recent computer display controllers will generate a 16:9 image. The two resolutions in this ratio listed in Table 1 are commonly described as HDTV, but resolution for television is complicated by the compression and encoding of the signal—see Poynton's book for more detail.

Display sizes are measured diagonally for a given aspect ratio; these are often more a label than an actual measurement. A 19″ monitor, for example, may have only a 17–18″ diagonal displayed image.

The size of a projected image depends on the projector optics, which define both the size of the image for a given distance from the projector (called the *throw distance*), and whether the image can be focused at that distance. Most projectors have a focus control that fine-tunes the focus at a given size. Many projectors have a zoom lens, which gives a wider range of sizes.

The most convenient way to specify projected image size is to use a *throw ratio*, which is the ratio of the throw distance to image width. This, plus the range of throw distances over which the image will focus, gives a complete specification for a known aspect ratio. For example, a projector with a fixed lens may have a throw ratio of 1.75:1 over a throw distance of 3′–30′, which means the image ranges from 1.7′–17′ wide. Outside of this range, the image will not focus. Unfortunately, the specification for projector image sizes is not very standardized. Some manufactures provide focal length information for their lenses rather than throw ratios. Others provide a throw ratio that is the ratio of the throw distance to the diagonal size of the image, not its width.

## Intensity Resolution

Intensity resolution would ideally describe the number of visible intensity steps produced by a display. However, what constitutes a visible step, as previously described in Chapters 2 and 3, is a complex matter depending on the color, absolute intensity, and viewing environment. Therefore, what is usually specified is the number of digital steps used to define the pixel intensity for red, green, and

(a)                  (b)

Figure 7.
Contour lines caused by insufficient intensity resolution. (a) Original; and (b) quantized to 4 bits/pixel.

blue. In the digital display world, this is almost uniformly encoded as 8 bits in order to produce 256 levels.

The distribution of intensities produced by these bits depends on the transfer function, which can be modified either in the display hardware or by the display controller hardware and software.

Insufficient intensity resolution will introduce visible quantization, or *contour lines*, as shown in Figure 7. Sufficient intensity resolution to avoid quantization depends on the total contrast of the display, as well as the distribution of the intensity values. The more perceptually uniform the distribution, the fewer steps are needed. While precise numbers are very context dependent, it is possible to give a few pragmatic ones. A display with the usual non-linear CRT transfer curve, which approximates a perceptually-uniform brightness distribution, rarely shows contouring at 8 bits/color, except in some critical situations. Linearly encoding on such a display needs at least 10 bits, with some experts arguing for 12 bits. Those producing digital imagery for movie film, which has higher contrast, may use a 10-bit logarithmic encoding, or a 16-bit linear one.

## Number of Colors

The number of colors achievable on an additive display is, in theory, the product of the intensity levels for each of red, green, and blue. Given 256 levels per primary, for example, the total number of colors is $256 \times 256 \times 256$, or approximately 16.78 million colors. This doesn't, however, define how many million visibly different colors there are, which depends on the same perceptual factors as intensity resolution, only more so. As a result, the number of colors quoted for a display is more accurately a technical description of the display controller than a perceptual description.

A display controller includes a *display memory* (or *frame buffer*, or *video memory*) that encodes the color for each pixel. These colors are decoded and output to the display hardware, usually as 8-bit signals for each primary. Different display modes use different numbers of bits to encode each pixel, either 8, 16, or 24.

Display controllers that independently store 24 bits for each color (8 bits/color/pixel) are called *full color*, or *true color*. Some true color controllers include a fourth channel to support transparency and blending in 3D graphics, which means that they store 32 bits/pixel. This is called an *alpha* channel (see Chapter 10).

Back when memory was expensive, it was common to encode colors to 8 bits/pixel and use a lookup table, called a *color map*, to translate from 8-bit pixel values to 24-bit color values. This is often called *indexed color*, as the pixels act as an index into the color map. For example, in the color map table, the pixel value 128 might be assigned to a triple of red, green, and blue values, such as (R: 45, G: 78, B: 92). Only the index value, 128, is stored in the frame buffer, but the full RGB triple is sent to the display hardware. Changing the color map will change the displayed color without changing the frame buffer contents, a technique that can be used for animation.

Using indexed color reduces the total number of colors to the number of index values, for example, 256 for an 8-bit index. Indexed color is still supported on most PC systems. Some animated games require indexed color for performance reasons, as it is faster to update 8 bits/pixel than 24 bits/pixel. There is also a standard indexed color scheme used in web design.

The display mode called *high color* stores 16 bits/pixel, allocated as 5 bits each of red and blue, plus 6 bits of green. Green is given the extra pixel because the eye is more sensitive to variations in green intensity. This mapping is intrinsic to high color, and cannot be changed. There are approximately 65,000 color combinations with this encoding, a significant improvement over 8-bit indexed color. Many users find high color functionally indistinguishable from true color, although some images will show quantization artifacts compared to true color.

When using fewer than 24 bits/color, colors not in the color map must be approximated with high-frequency patterns of color, a technique called *dithering*. Dithering can add visible texture to colors. Figure 8 compares true color, high color, and 8-bit indexed color. The image is a small star displayed on the screen and blown up large enough to see individual pixels. Any texturing in the true color mode is due to the printing process. The dither pattern is very obvious in the 8-bit mode, and should be just visible in high color mode, especially across the center of the star.

Independent of the display mode, display controllers often include lookup tables on the output side that change the transfer curve for each color. This is the way, for example, the operating system can globally change the transfer function for all applications. These tables traditionally map one 8-bit value into another, which means that in general, multiple pixels will map to the same color. Careless use of these tables can introduce quantization, even in true color mode.

## Additive Color Reproduction

Additive color reproduction means taking an image, which is encoded as additive RGB pixels, and reproducing it on an additive color output system. An alternate description is to take an image viewed on one display and make it "look the same" on another. This problem was described mathematically in Chapter 3 for two ideal color RGB systems as a sequence of table lookups (to map the brightness values) and matrix multiplications (to convert from one set of RGB primaries to another).

The same principles can be applied to physical displays if their color properties match those of an ideal RGB color space as described in Chapter 3. In an ideal RGB system, the colors of the

(a)

(b)

(c)

Figure 8.
Comparison of (a) true color; (b) high color; and (c) indexed color, which shows the effects of dithering.

primaries do not change with brightness, and each pixel's color is independent of the colors around it. To be able to use $3 \times 3$ matrix transformation, the light emitted at (R:0, G:0, B: 0) must be negligible. While CRT displays match this ideal model well, most new color display technologies deviate to some degree, some significantly.

It is not always necessary to perform a full three-dimensional mapping to create a good color reproduction. Many display systems have similar primary colors and vary only in the transfer function and white point. Simply matching the individual R, G, and B transfer functions will compensate for both of these effects. Even if the systems in question do not follow the ideal model, this can go a long way toward matching the appearance of two additive color systems.

If there is significant difference in the color of the primaries, some sort of global remapping will be needed. Even if the display deviates somewhat from the ideal RGB model (and many do), a matrix transformation can improve the color fidelity when the primary colors are significantly different.

Converting from one RGB gamut to another can create colors that fall outside the target gamut. Mathematically, these colors will contain values greater than 1.0 or less than 0. The most expedient way to handle these colors is simply to clip them into range, which can introduce hue and shading artifacts. More sophisticated gamut mapping algorithms will be discussed in Chapter 9, as part of color management systems.

Converting color images on a digital display to a video encoding includes both a color space transformation and further processing to encode and compress the signal. For any video standard, there is a specified RGB color space, such as SMPTE RP 145 (SMPTE-C) for commercial video systems, or ITU-R BT.709 for high-definition television. A first step for achieving color fidelity is to transform into this space, but the compression used in video may affect the quality of the color reproduction. Analog video has not only significantly reduced spatial and intensity resolution for the color signal—its encoding introduces interpixel color effects, such as the inability to make sudden hue shifts in certain parts of the color space. These problems are less apparent in digital video, although color artifacts due to compression may still be visible.

# CRT Displays

CRT monitors excite phosphors with rays of electrons emitted from an *electron gun* to create colored light. The red, green, and blue phosphor dots are arranged in a fine pattern on the screen surface, as shown in Figure 1. Many phosphor dots contribute to a single pixel (Figure 9), which is defined by the size and position of the electron beam. Since there is no physical structure defining the spatial resolution other than the fine spacing of the individual phosphor dots, a monitor can smoothly operate at a wide range of resolutions.

The phosphors coat the inside surface of the display. There are three electron guns operating simultaneously in a color CRT display, one per color. The *shadow mask,* a fine grid of holes or slots positioned immediately behind the phosphors, helps constrain each beam of electrons to excite only the desired phosphor color. Variations in brightness are achieved by varying the power of the electron beam.

The phosphor colors, which define the color gamut, are essentially the same for all modern monitors, and match the sRGB gamut (Figure 4). As CRTs age, however, they become progressively less bright, and usually somewhat yellowish in color. This is due to the deterioration of the phosphors, with the blue degrading faster than the other two colors. This does not change the position of the primaries on the chromaticity diagram, simply the balance between the red, green, and blue colors (the blue is less bright, making the overall appearance more yellow). Up to a point, it is possible to compensate by adjusting the brightness and contrast controls for each primary independently.

Most CRT displays used on computer systems include hardware controls to set the white point to one of a set of standard values, such as D65; these, however, are simply the correct settings for a new monitor. Unless the display includes some form of light measuring instrument, like the Barco Calibrator line of displays for the graphic arts, there is no way for the electronics to determine what light is actually displayed. Therefore, as a CRT ages, these standard settings will become progressively less accurate. A rule of thumb is

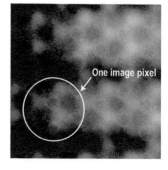

One image pixel

Figure 9.

Each pixel on a CRT display is a cluster of phosphor dots.

that two years of continuous use will cause visually significant aging in the color and brightness of a CRT display.

A CRT's native transfer function is a non-linear power function, which is defined primarily by the physics of an electron gun exciting phosphors. In its simplest form, it is approximated by

$$I = V^\gamma,$$

where I is the measured intensity and V is the input voltage, which corresponds to the input pixel values. If the intensity produced at $V_0$ (black) is not negligible, then it must be included in the equation as follows:

$$I = (V + V_0)^\gamma .$$

The entire curve can be scaled by a constant, k:

$$I = k(V + V_0)^\gamma .$$

The effect of the brightness and contrast controls described above and illustrated in Figure 5 is to change $V_0$ and k, respectively.

To get the best image on a CRT, the usual recommendation is to adjust the contrast to the highest value it can go without distorting the white pixels (called "blooming"), and the brightness so that no light is emitted at $V_0$. For most monitors, the exponent, $\gamma$, lies between 2.2 and 2.5. Exactly how to measure and specify gamma for CRT displays is, surprisingly, still a topic of some debate. A reasonable interpretation is that the electronics produce a value of 2.5, but that the net system effect, including both physical and perceptual factors, is closer to 2.2.

CRTs can readily be modeled as an ideal RGB system, as long as the non-linear pixel-to-luminance brightness curves are correctly characterized for each primary.

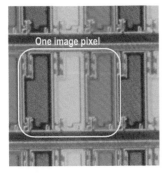

One image pixel

# Liquid Crystal Displays

Liquid Crystal Displays (LCDs) are spatial arrays of red, green, and blue segments, as shown in Figure 10. Each of these segments is a colored filter over a cell of liquid crystal material that can be made

150

variably transparent. A backlight shines through the LCD array so the resulting color is a function of both the filters and the backlight.

In LCDs, a single pixel is one triple of red, green, and blue elements (or it may contain two green elements), which gives the panel a natural spatial resolution. Operating it at other resolutions can create images that appear jaggy—this is generally true for all flat panel displays.

Colored filters are quite different from colored phosphors: Instead of converting a high-energy electron beam to visible light, color is filtered from the white backlight. The more intensely colored the filter, the less light it will pass, making the display less bright. A less saturated filter would create a brighter but less colorful display given the same backlight. The only way to have a bright, highly colorful display is to have highly saturated filters and a very bright backlight. This consumes more power, and therein lies the trade-off, especially for battery-operated displays.

Most users are accustomed to the highly-saturated primaries of a CRT display, which forces all display manufacturers towards that "standard." At this writing, the red and green primaries of a desktop LCD tend to match those of a CRT in color, but the blue is still less saturated. The display on a laptop computer, however, is significantly less saturated in all colors than a CRT or desktop system; this is shown on the chromaticity diagram in Figure 11, which compares two desktop LCDs and a laptop LCD to a CRT (sRGB) gamut.

The color of LCD panels change with viewing angle, some-

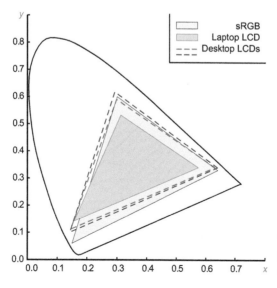

Figure 11.

Chromaticity diagram comparing the gamuts of a laptop LCD and two desktop LCDs to the sRGB gamut.

151

times to the point of generating color and brightness inversions. This has improved dramatically in the last few years, especially for desktop LCD systems. In contrast, light scatters almost uniformly from the surface of a CRT display.

The native transfer function of an LCD is roughly linear, with roll-off at each end. However, LCD displays usually include electronics so that their transfer functions mimic monitors to achieve compatibility with existing systems.

The white point of a LCD is defined by the filters and the backlight. There is a natural white point, defined by the backlight shining fully through the filters. This is the white point seen on laptop displays, for example. Desktop displays, however, include white point menus like CRTs, similarly changing the balance of the red, green, and blue light. Color changes due to aging are primarily caused by aging in the backlight, rather than aging in the filters.

LCDs emit visible light at $V_0$ due to the backlight leaking through the front of the display, in contrast to CRTs, whose black emissions are usually negligible. This is sometimes called *flare*. Flare adds a constant level of light to all colors produced by the display, and as a result, changes the color of the primaries, rather than having them be constant for all brightness levels, as in the ideal model. This is shown in Figure 12(a), which are measurements on an Apple Cin-

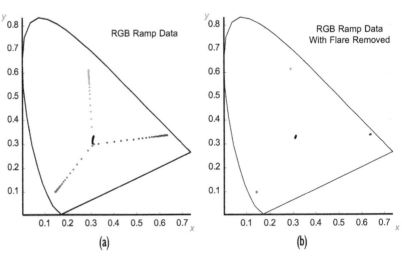

Figure 12.

The effect of flare light on the chromaticity of the primaries of an Apple Cinema LCD. (a) With flare; (b) with flare subtracted. (Figure courtesy of Jason Babcock.)

152

ema display. Each dot represents a measurement at a different brightness level. The chromaticities shift toward the color of "black" as they get darker, rather than remaining constant. Subtracting the flare light from all measurements corrects this problem, as shown in Figure 12(b). For color characterization, subtracting the flare light is represented most compactly as a $4 \times 4$ homogeneous matrix, as described in Chapter 3.

Even with flare subtracted, LCDs deviate measurably from an ideal RGB model because the chromaticity is still not constant at all brightness levels. If it were, the cluster of points at each primary color in Figure 12(b) would map to a single point. Note especially the drift from white to black (middle point). Whether this variation, which is primarily in the darker colors, is significant depends on the display and the application.

# Plasma Displays

Plasma displays create light by sparking a plasma discharge that emits ultraviolet light, which then excites phosphors that coat the plasma cell surface. This produces bright, saturated colors that are not subject to the sort of viewing angle problems that affect LCD systems. Grayscale is achieved by creating a high-speed sequence of light pulses. This creates a naturally linear transfer function, which is electronically modified to match the classic CRT transfer function. Like LCDs, plasma displays are spatial arrays of red, green, and blue segments, and one triple or quadruple/pixel, which defines their ideal resolution.

Some of the largest, most dramatically bright and colorful flat-panel displays use plasma technology. Compared to LCDs, however, they are relatively low-resolution, high-powered, and expensive. They also age relatively rapidly, due to the action of the plasma discharge on the phosphors. Aging makes the pixels dimmer or makes them fail entirely.

Unlike LCDs, the primary colors for plasma displays do not seem designed to match that of a CRT display. This is probably because the production of plasma displays is directed towards home enter-

Figure 13.

The gamut of a Pioneer plasma display compared to the sRGB gamut, and the video gamuts, NTSC, and SMPTE-C.

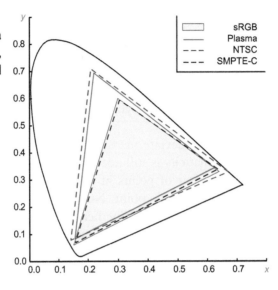

Figure 13.

The gamut of a Pioneer plasma display compared to the sRGB gamut, and the video gamuts, NTSC, and SMPTE-C.

tainment systems rather than computer displays. Figure 13 shows a plasma display gamut plus the sRGB gamut (equivalent to the HDTV gamut, ITU-R BT. 709), and two analog video gamuts, NTSC and SMPTE-C. While still visibly different, this particular plasma display gamut (a Pioneer PDP-503CMX) is most similar to the old NTSC color space.

Plasma displays can contain special electronics that limit their brightness to extend their lifetime. This means that colors will grow monotonically brighter up to a point, then will be clamped to some maximum value. Operating in this mode, plasma panels clearly do not model ideal additive systems.

If kept in a mode that avoids this clamping, plasma displays can be characterized with the ideal model. Like LCDs, they exhibit flare that must be subtracted to achieve an ideal model. The existence of this flare light suggests that the plasma cells are never entirely turned off, as there is no other source of light in the system.

## Projection Displays

The first video projectors projected the light from three, bright monochrome CRTs through red, green, and blue filters. These three, physically separate images were painstakingly overlapped and aligned (converged) to create a single, full-color image, similar to the diagram in Figure 2. Such systems still exist for specialized applications such as display walls, and include sophisticated electronics and controls for aligning and blending the images.

A digital projector contains a digital imaging element such as a small LCD or an array of micro-mirrors (Digital Micro Display, or DMD) that modulate the light from a high-intensity light bulb. Most LCD projectors and the larger DMD projectors contain three imaging elements and a dichroic mirror that splits the white light from the bulb into its red, green, and blue components. These are recombined and displayed simultaneously through a single lens. The quality of the optics for the splitting and recombining strongly affects aspects of the image quality: color, uniformity, sharpness and convergence. There are many different complex mechanical/optical configurations for projectors—see the Stupp and Brennesholtz book for the details.

The smallest projectors use a single imaging element and a color wheel of filters, so the separations are displayed sequentially. The DMD projectors based on the Digital Light Processing (DLP) technology from Texas Instruments can weigh less than 3 pounds and still produce a bright, crisp, image. Their color wheel includes a clear filter that is used to add white light to the brighter colors. This increases the brightness and contrast of the system, but only for a limited set of colors near white.

The color for a projection system is defined by the bulb color plus the filter colors. There is variation in both, as well as variation in the bulb due to aging. Identical projectors often have visibly different color gamuts, although this can be minimized by carefully matching the R, G, and B transfer functions.

The primary colors chosen for digital projectors are often visibly different than those for displays, creating substantial color shifts when colors designed on a display are projected. Figure 14 shows the primaries for five different digital projectors plotted together with the gamuts of an sRGB CRT and a laptop LCD display.

The native resolution for a projector is defined by its imaging element. Displaying other resolutions is implemented by electronically resampling the input image. In some projectors, especially older projectors (3–5 years ago), this resampling is very poor, rendering text

Figure 14.

The gamuts of five different digital projectors compared to the sRGB gamut, and that of a laptop display.

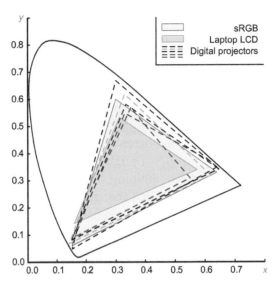

illegible. More modern systems do a substantially better job.

The native transfer function for a projector is defined by the imaging element. DMD elements pulse to encode the grayscale, making them naturally linear, whereas LCD elements are identical to liquid crystal displays. However, like flatpanel displays, most projectors contain image processing hardware that induces a transfer curve compatible with CRTs and video. They also have similar brightness and contrast controls, though they may only be visible through menus accessed with the projector's remote controller.

A CRT projector, like a CRT display, can be described as an ideal additive RGB system. Three-element DMD projectors exhibit flare, but can be characterized with a 4 × 4 matrix. LCD projectors exhibit both flare and the same deviation from the ideal color model near black seen in LCD displays. The single element DMD projectors, with their added white light, cannot be characterized with a simple additive model.

## Characterizing Additive Color Systems

To characterize an additive system means to create a transformation between the RGB pixel values and a visual color space such as CIE XYZ or CIELAB. Just as RGB pixels are fundamental to digital color, being able to specify these pixels with respect to some metric is fundamental to controlling digital color.

156

The introduction of color management systems at the consumer level has created a market for relatively low-cost display characterization hardware and software. Commercially, characterization data is called a *profile*, which is the terminology used in color management systems (Chapter 9). A package with both a display colorimeter and software to drive it costs well under $500. It is also possible to construct an approximate characterization from a combination of manufacturer-supplied information about the chromaticity of the primary colors, and visual estimations of the transfer curve and white point.

Before starting to measure, it is important to use some form of process control to establish a standard setup and ensure repeatable results. For additive systems, this generally means establishing a smooth, effective transfer function, as well as identifying and controlling all parts of the system that affect color and brightness. It also means setting the white point. Once the desired settings are established, it is important to disable all controls to avoid accidents. It is common to see critical display systems with tape over the controls.

Most additive systems vary significantly in brightness across the displayed image. A CRT may be 30% less bright in the corners than the center—projection systems can be worse, though good quality ones are much better. For a projection system, this variation is affected both by the optics and by the projection screen. Variation in LCD displays is a function of the backlight, whose variations may be smaller yet more visible than those on CRTs because they change less smoothly. Viewers are relatively insensitive to these variations in brightness, but instruments are not. Therefore, it is important to sample in exactly the same position each time. Characterization software creates a target to indicate where to place the instrument, which is almost always pressed onto the surface of the display.

Commercial characterization systems are tuned to a very specific task: creating data files for color management systems. The typical customer of such a system is a professional graphic artist, so the controls (and technical descriptions) provided are minimal. Often, such systems do more than measure; they routinely set the display controller's lookup tables to set the transfer curve to a specific function, usually a specific value of gamma. Their only output is data files in the profile format used in color management. There is no way to control what they produce, or even a convenient way to read it, making these systems difficult to use for other purposes than creating profiles.

Those interested in characterizing additive color display systems for experimental work need, at minimum, a display colorimeter and software to control it. These need to be coupled with software that will display suitable colored patches. It is important to be sure that the instrument will accurately measure the display spectra. For example, the red CRT phosphor has several narrow spikes. Some instruments will miss these spikes and underestimate the brightness of red. Also, display measurement equipment designed for CRT displays may not be accurate on other displays, as they contain filters optimized for CRT phosphors. Check before you buy.

The key constraint for using the ideal RGB model for characterizing an additive system is that the primaries display a constant chromaticity independent of their brightness. This will be guaranteed if the spectral distribution of each primary is simply scaled as the brightness decreases, which is the test applied in most technical papers on characterizing displays. Measuring chromaticity is equally valid, but potentially more sensitive to error introduced by the measuring equipment.

To use a $3 \times 3$ matrix to transform the primaries, the light emitted at black should be negligible. If there is significant light emitted at black (flare), it must be subtracted first, as shown in Figure 12. As mentioned above, CRT displays and projectors have sufficiently low black levels to ignore flare for most applications. LCD displays and projectors exhibit both flare and cross-talk between the primaries at low brightness level that can affect the predicted value of dark colors. DMD projectors with three imaging elements will match the ideal model once the flare is subtracted. Flare can be subtracted in the matrix by using a $4 \times 4$ homogeneous matrix, as described in Chapter 3.

Another key assumption in the ideal RGB model is that each pixel is independent; that is, the color of the surrounding pixels does not affect the color of the measured pixel. This is rarely precisely true, due to either electronic loading, or light leaking from surrounding cells. It is simple to test for spatial independence by measuring with a variety of different background colors.

As was mentioned for both small DLP projectors and plasma displays, some display systems deviate significantly from a pure additive RGB model. The DLP projectors add white in steps; the plasma displays clamp brightness to increase lifetime. A quick test for this sort of behavior is to compare the colors measured along the gray axis to the additive sum of the primary colors set to the same pixel levels. Full white on a small DLP projector, for example, is 43% brighter than the sum of the primary colors, a strong hint that something unusual is going on.

Rather than its use a model, it is also possible to use a 3D lookup table to characterize an additive display system. In this case, measurements are taken throughout the color space, and interpolation is used to estimate intermediate colors. This eliminates the need for each primary to have constant chromaticity, but does not compensate for variations caused by adjacent pixels. While such an approach is more common for subtractive systems, nothing precludes using it for display systems also. Some commercial display characterization systems do this, especially for LCDs.

## Summary

Digital color displays attached to computers escaped the research lab in the early 80s as components of personal computers and video games. Now, they are a ubiquitous part of the computer interface, and have fundamentally changed the nature of computing and its applications. They have also fundamentally changed the nature of games, a development less universally applauded.

Even five years ago, most digital color displays used cathode ray tubes (CRTs), the same technology ubiquitous in television. Now, high-quality flat-panel displays are becoming ubiquitous, along with small, bright, and affordable digital projectors. Coupled with a laptop computer, digital projectors are making slide and overhead projectors obsolete.

Additive color systems, such as displays and digital projectors, render the RGB components of a digital color image by transforming from the image primary colors to the output primaries, then carefully reproducing the correct brightness values for each primary. As long as the original and the output contain a similar set of colors, the result will be pleasing and accurate. If the gamut of the original image and the output device are quite different, then an exact match is impossible, and a pleasing result requires more art.

Anyone involved in the precise computation of RGB pixels should have a precise specification of their color and brightness. Commercial display characterization systems can easily be used to ensure consistency, and to provide metrics that make reproducing computer-generated imagery more reliable.

Those interested in measuring and characterizing additive displays for research or engineering purposes will need to create their own characterization data. Those most particular about display characterization are those involved in research on color vision. The contribution by David Brainard, et. al. in *The Encyclopedia of Imaging Science and Technology* is an excellent description of display characterization in general, and the precision needed to apply these techniques to vision research. William Cowan's 1983 article, which was published at SIGGRAPH 83, is a good reference that is readily available to the computer science community. Roy Bern's 1996 article in *Color Research and Application* is a standard reference in the digital color research community.

Compensating for differences between RGB gamuts is well within the capability of modern graphics hardware, as it is fundamentally table lookups and matrix multiplications. In many situations, simply matching the RGB transfer functions between systems, which is all table lookup, can go a long way toward making two additive reproduction systems "look the same." The proliferation of different additive color systems makes color matching across them an issue of practical importance, as anyone who has had presentation colors designed on a laptop shift into illegibility when projected can attest. In the future, accurate RGB gamut transformations should become a standard part of any color reproduction and rendering system.

References.

R.S. Berns. "Methods for characterizing CRT Displays." *Color Research and Application* 16 (1996), 173–182.

W.B. Cowan. "An Inexpensive Scheme for Calibration of a Colour Monitor in Terms of CIE Standard Coordinates." *Computer Graphics (Proc. SIGGRAPH '83)* 17:3 (1983), 315–322.

# 7  Additive Color Systems

# 8
# Subtractive Color Systems

T ake a photograph or print a page. Light shining through layers of colored dyes or inks creates the color image. Printing inks and photographic dyes selectively subtract the red, green, and blue image components from white light, hence the name "subtractive color." While some subtractive color images are projected, such as slides, most are prints, which can be viewed in any lighting. Inexpensive color prints are the ultimate value provided by subtractive color. This is the fourth chapter on image reproduction technology; it describes the basic principles that underlie all subtractive reproduction, some technology-specific details, and models for characterizing subtractive systems to device-independent color standards.

## Introduction

Subtractive color systems put colored images on paper, either by printing with ink or by exposing special photographic paper. Projected slides and movies are also examples of subtractive color reproduction.

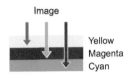

Figure 1.

Film layers selectively absorb red, green, and blue wavelengths to capture an image.

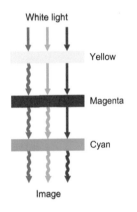

Figure 2.

An image is reproduced by selective filtering a white light source. The cyan, magenta, and yellow layers recreate the red, green, and blue components of the image.

Printing and photography, the subtractive color industries, have made color reproduction a ubiquitous part of modern life.

Subtractive color uses layers of cyan, magenta, and yellow filters (inks or dyes) to create color. In photography, the dye layers in the film sequentially absorb blue, green, and red light, as shown in Figure 1. When developed, white light filtered through the layers recreates the image, as shown in Figure 2, which illustrates the process for a projected slide. Each layer (ideally) modulates one of the red, green, or blue components of the white light source, as indicated by the wavy lines. The other components are left unchanged. These layers, plus a single white light source, are what define a subtractive color system.

Prints on paper filter the light twice, as shown in Figure 3, which illustrates the principle for a single colored layer (either printing ink or a dye layer in photographic paper). White light passes through the colored material, reflects off the paper, and is emitted as colored light. A full-color print would have three or four ink layers; in printing, black ink is added to increase contrast. The smoother and whiter the paper, the more light is reflected, which is why prints always look better on bright white coated papers.

Subtractive color creates color by selective absorption, and therefore depends critically on the spectral distribution, not just the perceived color, of the illuminating light. Change the lighting and the color changes, which is why controlling the lighting is critical when evaluating images created with subtractive color processes.

The principles and processes of photography and printing were originally created using optics and chemistry. Over the last 20 years, digital technology has put color printing and production on every desktop, and recreated the graphic arts industry. A similar revolution is changing photography and cinematography. Digital cameras and printers are converting home computers to consumer-level digital darkrooms. DVDs, digital special effects, and ultimately, digital cinema, are making digital color imaging an intrinsic part of modern cinema.

Digital printers and film recorders are the technologies that transform RGB pixels into subtractive color. A digital printer layers ink or toner on paper, either spraying ink, fusing toner, or digitally creating plates for printing. A digital film recorder exposes film from a digital image. The film is chemically processed in the usual way to create a print or a transparency.

Compared to additive color systems, creating pictures with subtractive color is a slow, complicated, variable, and often mysterious process. Reading this chapter should make subtractive color less mysterious, and in doing so, may make it somewhat easier to use. It will not, however, provide any "silver bullets" for making prints automatically match displayed color—this chapter can only help describe why this is an inherently unsolvable problem. Color management systems, which are the topic of the next chapter, provide a framework for more accurate transformations from RGB to CMY(K).

Photography and printing are two large, well-established fields. There are literally hundreds of books on these topics. Hunt's *Principles of Colour Reproduction* includes a good description of both printing and photography as traditionally practiced, as well as television. Phil Green's *Understanding Digital Color* is a better reference for the practitioner. In the printing area, one of the most respected references is John Yule's book, *Principles of Color Reproduction,* which was recently revised by Gary Field and reissued (the original version is out of print). The *Focal Encyclopedia of Photography* is a complete, if rather terse, reference for the technology of photography and film. *Basic Photograph Materials and Processes,* by Stroebel et al., is a good textbook for the technical side of photography. Other references are provided in the bibliography. The Society for Imaging Science and Technology's (IS&T) conferences and publications are the standard venue for research in photography and printing, especially the digital forms of these technologies.

This chapter begins with a discussion of the basic principles of subtractive color, described in terms of ideal CMY filters. This is followed by a discussion of how the less-than-ideal filters used in real color technology influence this model, including the addition of black ink in printing. The next sections discuss the two ways subtractive systems control variations in lightness: density and halftoning. This is followed by a discussion of film recorders and

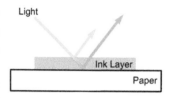

Figure 3.

In a color print, the light passes through the color filter layer (ink or dye) twice, reflecting off the surface of the paper.

printers and the technology of subtractive color reproduction. The chapter concludes with a summary of subtractive color reproduction and how device-independent color is applied to such systems.

# The Theory of Subtractive Color: Ideal CMY Filters

The inks and dyes used in subtractive systems act as colored filters. Each is designed to independently modulate distinct red, green, and blue components of the spectrum, and leave the rest of the spectrum untouched. The filter that passes green and blue but modulates red appears cyan. Similarly, the green-modulating filter appears a purplish-red called magenta, and the blue-modulating filter looks yellow. Ideal versions of these filters, called *block filters*, are shown in Figure 4, which shows a spectrum divided into its red, green, and blue components, plus the corresponding block filters. These block filters absorb 100% of their associated RGB component.

The CMY color wheel is shown in Figure 5, which includes the secondary (pairwise combination) colors, red, green, and blue. While they have the same names, these are not the vivid primary colors of a light-emitting, additive system. Overlaying all three primary colors (ideally) subtracts out all of the light, and produces black. Black ink, which is used in most printing systems to improve the overall contrast, is not an independent primary color.

Figure 6 shows how to create color (R: 0.75; G: 0.5; B: 0.25) with ideal block filters. White light, which is 100% red, green and blue, enters from the top. The yellow filter subtracts (absorbs) 75% of the blue light; the magenta filter subtracts 50% of the green light; and the cyan filter subtracts 25% of the red light.

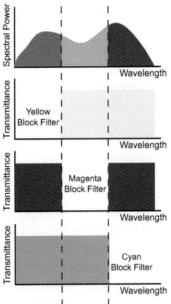

Figure 4.

Ideal subtractive color filters are block filters that absorb only one band of the spectrum and fully transmit the rest of the wavelengths.

The light exiting at the bottom, therefore, is 75% red, 50% green, and 25% blue, as desired.

This example illustrates several important points about subtractive color. In this ideal model, the absorption of the filters is the inverse of the intensity of the desired light. When CMY is specified as digital color, 1.0 means full absorption, as was shown in Figure 4. As a result:

(C: 0.25, M: 0.5, Y: 0.75) = (1, 1, 1) − (R: 0.75, G: 0.5, B: 0.25) .

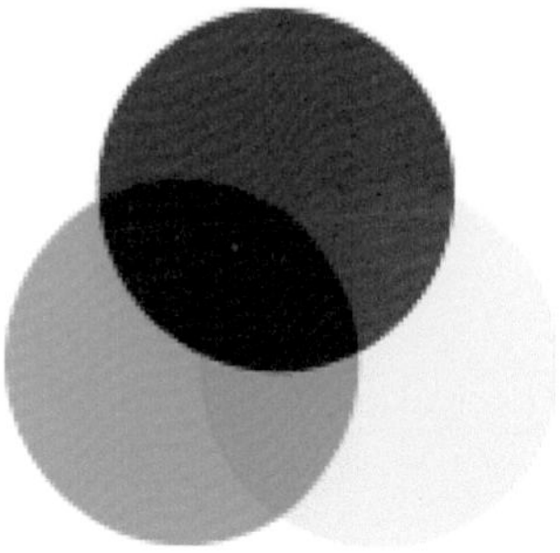

It's important to emphasize that this simple model only "works" for ideal block filters, which independently modulate the three parts of the spectrum. If the filter functions overlap, as is commonly the case, then the three components of the outgoing light cannot be computed independently—they depend on the absorptance of 2 or 3 filters. This will be discussed further in the next sections, which discuss the realities of non-ideal CMY(K) colors.

Different technologies have different ways of varying the absorption of the filters. In photography and some printing systems, the filter layer becomes more or less opaque, either by varying its thickness, or through the film development process. In printing, it is more common to create a pattern of full-color dots, whose coverage determines the amount of absorption. How printers vary brightness, including the use of density measure-

Figure 5.

The primary colors of a subtractive color system and their combinations.

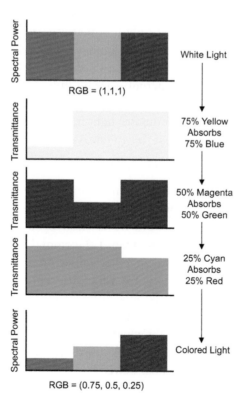

Figure 6.

The process of creating color with ideal layered block filters.

167

ments and the addition of black, is described in more detail later in the chapter.

The "RGB" used in this model are the colors produced by lights whose spectral distributions are the inverse of the block filters, or more precisely, the product of the light source with the inverse of the block filters. Unlike additive color, where the red, green, and blue color values act as tristimulus values, the precise spectral distributions for RGB are important in subtractive color systems. This is why subtractive color cannot be characterized with a simple, trichromatic model like additive color.

## Real Subtractive Color: Non-Ideal CMY

The cyan, magenta, and yellow filters used in subtractive media are never as ideal as described in the previous section. The cyan filter, which should only modulate the red part of the spectrum, also affects the green part. Similarly, magenta modulates both red and blue as well as green. Figure 7 shows a comparison between ideal "block filters" and actual printing inks. With non-ideal filters, there is interaction, or cross-talk between the filters. The magenta filter, for example, modulates light in both the red and blue parts of the spectrum, not just the green. As a result, in real systems:

$$CMY \neq (1, 1, 1) - RGB .$$

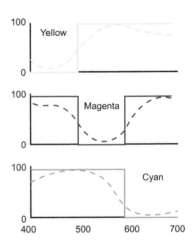

To emphasize this point, compare the patches in Figure 8, which are what would be created by a simple "subtract from 1.0 formula" to the equivalent pixel values on a monitor—they are very different. The most strikingly different color is probably the blue. On the monitor, it is a bright, vivid color, whereas on print it is dark, and nearly purple. A much better blue can be created as (C: 1, M: 0.5, Y: 0), as is

Figure 7.

Real colored filters (for printing ink) contrasted with ideal block filters.

168

also shown in the figure, along with a better green; red is unchanged. Cyan on a monitor is a very bright blue-green color, whereas printing cyan is much more blue, and the monitor magenta is a bright, vivid pink compared to the darker, more reddish printer's ink.

The effect of the ink impurities is especially visible in the reproduction of gray. In real subtractive color systems, equal amounts of the primary colors rarely yields gray. For example, Figure 9 shows two gray patches, one printed with equal amounts of CMY, and the other, with the amounts of CMY carefully balanced to produce a neutral gray (there is no black in this patch). In the balanced patch, there is significantly less magenta and yellow than cyan. This balance will vary over the grayscale, which makes creating a neutral grayscale one of the challenges of subtractive color.

Film dyes also deviate from ideal block filters, though less so than printing inks. Modern films may also have additional dye layers whose function is to filter out the unwanted cross-coupling. Like print, creating a neutral grayscale on film requires balancing the colors, from exposure through processing. In a film recorder, creating a neutral grayscale for a particular film and its processing is one of the fundamental calibration steps.

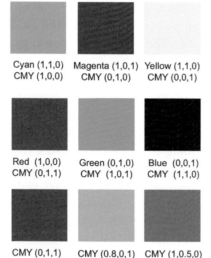

Cyan (1,1,0)  Magenta (1,0,1)  Yellow (1,1,0)
CMY (1,0,0)   CMY (0,1,0)      CMY (0,0,1)

Red  (1,0,0)  Green (0,1,0)   Blue  (0,0,1)
CMY (0,1,1)   CMY  (1,0,1)    CMY  (1,1,0)

CMY (0,1,1)   CMY (0.8,0,1)   CMY (1,0.5,0)

Figure 8.
The colors produced by simply inverting the RGB pixel representation to get CMY values for red, green, and blue. The bottom row shows the "standard" red, green, and blue used in the illustrations in this book.

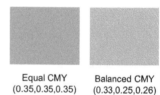

Equal CMY        Balanced CMY
(0.35,0.35,0.35) (0.33,0.25,0.26)

Figure 9.
Equal CMY does not create a neutral gray. To create a three-color 35% gray requires a balanced blend of CMY, as shown. Darker colors would also include black.

## Adding Black: CMYK

In color printing, black ink is added primarily to improve contrast. It is also used to replace expensive colored ink with cheaper black ink, and to prevent the paper from getting too wet. Black, in this context, is called K, for *key*, an old printing term.

Black is not an independent primary color like cyan, magenta, and yellow. It expands the color gamut, but only by making the dark colors darker. In typical printing practice, black both replaces and augments the CMY primaries. For any color that contains all of cyan, magenta, and yellow, there is a *gray component* equal to the minimum of the three colors. For example, the color (C: 1, M: 0.5, Y: 0.25) has a gray component of 0.25. This gray component is replaced with a mixture of cyan, magenta, yellow and black. Determining the amount of black to apply and the amount of CMY to remove is called *gray component replacement*, or GCR. An older term is *undercolor removal*, or UCR. In some contexts, UCR means adding black only for neutral colors, whereas GCR applies to all colors that have a gray component.

Figure 10 compares three sets of gray patches printed in black ink only and as balanced amounts of CMY plus black. There is no black in the 25% patch, a slight amount at 50%, and a significant amount at 75%. Both the balanced and black-only patches should appear neutral, although the balanced patches may not necessarily be exactly the same shade of gray as the black-only patches. The actual amounts of CMYK in the four-color patches are shown in the figure.

Figure 10.
Three steps of gray printed in black ink only compared to balanced amounts of CMYK.

25%    50%    75%

Black Only

Balanced CMYK

(25, 20, 20, 0)  (49,41,40,2)  (74,68,61,32)  CMYK %

Maximum contrast would be achieved by only adding black, but for many printing processes, printing full CMYK for black (called a 400% black) will make the paper so wet it buckles or tears, or may cause smearing because the ink cannot dry before the next page is deposited on it. Replacing some of the colored ink with black not only allows the paper to dry more quickly, but also saves cost.

Full GCR would replace all of the gray component with black. This is rarely done for printing presses, as the total contrast would be unsatisfactory. Figure 11 compares a black-only patch to a four-color black, which should look darker. In addition, misalignment

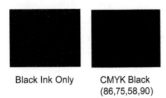

Black Ink Only      CMYK Black
                    (86,75,58,90)

Figure 11.
A black-only patch and a four-color patch, whose CMYK ink percentages are shown. The four-color black should look darker.

between the separations could introduce gaps between gray and non-gray colors. However, some digital printers, especially inkjet printers, have such difficulty with the paper getting too wet that full or nearly full GCR is used. Fortunately, registration on digital printers is accurate enough to allow this.

The process of defining the black component in printing starts with creating a neutral grayscale. The printing term *gray balancing* is the process of creating a neutral blend of cyan, magenta, and yellow for every step along the gray axis. Generally starting at around 50%, black is added to the mix. Both the neutrality and the perceived lightness of the patches must be adjusted to create a smooth grayscale.

There is no simple mathematical formula that predicts how to balance CMY to create a neutral gray, or how much black ink is needed to replace a specific blend of CMY, or to compute how much darker adding black will make a printed color—all of these depend on the specific inks, paper, printing process, and the order in which the separations are printed. The traditional solution is to plot a large number of patches with nearly equal CMY values, search for the grayest ones, and then iterate until a good grayscale is found. This was traditionally done visually, but now instrumentation is inexpensive enough that the process can be made more automatic.

# Lightness and Density

In subtractive color reproduction, variations in lightness are produced by varying the amount of light absorbed by the dye layers. For a given dye, the greater the absorption of the unwanted colors, the darker and more saturated the resulting color. For neutral colors, differences in absorption form the grayscale from black to white.

Absorption can be measured by comparing the incident to transmitted or reflected light. *Transmittance* is the ratio of these two quantities, expressed as a fraction or a percentage. A similar term is *reflectance*, which means the total light reflected off an object. *Opacity* is the inverse of transmittance.

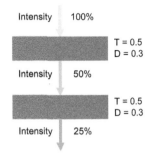

Intensity    100%

T = 0.5
D = 0.3

Intensity    50%

T = 0.5
D = 0.3

Intensity    25%

Figure 12.

Each 50% filter removes half of the light, leaving 25% of the original value. Equivalently, each filter has a density value of 0.3, which sum to a total value of 0.6, or 25% transmittance.

Combining two filters multiplies their transmittance. For example, two 50% filters reduce the light to 25% of its original brightness $(0.5 \times 0.5 = 0.25)$, as shown in Figure 12.

Density, which is a logarithmic measure of transmittance or reflectance, can be used to describe filter characteristics. Recall from Chapter 1 that density is defined as $-\log_{10}(T)$, or equivalently, $\log_{10}(1/T)$. A filter whose transmittance is 0.5 has a density of 0.3; a transmittance of 0.25 gives a density of 0.6. Because densities are logarithms, combining two filters sums their density values $(0.3 + 0.3 = 0.6)$, also shown in Figure 12. Density measurements, either transmissive or reflective, are commonly used for control and evaluation in both print and photography.

In subtractive color reproduction industries, it is common to use log-log plots to analyze tone reproduction, as shown in Figure 13. The x-axis represents the brightness values of the original image, either as the log of the normalized scene luminance or as density, and the y-axis represents the reproduced values as densities. The slope of such a plot is called its *gamma*. A gamma of 1.0 is a linear, 1:1 reproduction. Often, however, the reproduction cannot achieve the same density range as the original, as shown by the dotted line. Maximum print densities are usually no greater than 3.0; high-contrast transparency film densities can reach a value near 5.0.

Changing the gamma is the same as raising the input to a power equal to the slope. A gamma greater than 1.0 will increase the perceived contrast by compressing the dark colors more than the mid-tones and brighter colors, as was suggested by Bartelson and Brenneman to improve the appearance of reproduced imagery (see Chapter 2).

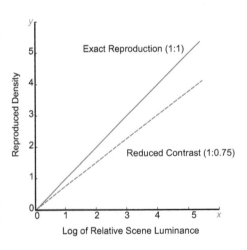

Figure 13.

Log/log tone reproduction curve, showing both a 1:1 reproduction, and one with reduced contrast.

A great deal of the art of subtractive color reproduction is compensating for ink impurities. One approach often found in traditional printing systems is to apply a $3 \times 3$ matrix to the color values expressed as density values to remove the unwanted contributions from the impurities. The matrix takes the following form:

Reference.
Michael G. Lamming and Warren L. Rhodes. "A Simple Method for Improved Color Printing of Monitor Images." *ACM Transactions on Graphics* 9:4 (1990), 345–375.

$$
\begin{aligned}
D_C &= a_{11}D_R + a_{12}D_G + a_{13}D_B \\
D_M &= a_{21}D_R + a_{22}D_G + a_{23}D_B \\
D_Y &= a_{31}D_R + a_{32}D_G + a_{33}D_B,
\end{aligned}
$$

where $D_R$, $D_G$, and $D_B$ are the desired RGB values, expressed as $\log_{10}(\text{intensity})$, and $D_C$, $D_M$, and $D_Y$ are the densities of the cyan, magenta, and yellow inks. The coefficients, $a_{11}$, $a_{12}$, etc., are chosen to create the corrected colors. If the inks were ideal, the diagonal terms would all be 1, and the off-diagonal terms zero. This approach is applied to printing on digital printers in a paper by Lamming and Rhodes.

The next section will discuss halftoning, which is the way printing systems create variations in lightness.

## Halftoning and Dot Area

Film creates variations in lightness by varying the opacity at each pixel. Most printing systems, however, are binary—ink either is printed or not. To get variations in lightness, the ink is printed in high-resolution patterns called *halftone patterns,* as shown in Figure 14. These patterns traditionally contain dots of varying sizes; the larger the dots, the more ink coverage and the darker the color. Halftone screens are specified as lines/inch, with 150–200 lpi typical for commercial printing.

Each screened separation is overlaid at a different angle, as can also be seen in Figure 14. This reduces the moiré patterns that can occur when two high-frequency patterns are overlaid and slightly

Figure 14.
Halftone patterns, as are used in offset printing. Each screen is rotated with respect to the others to minimize moiré patterns.

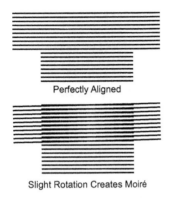

Perfectly Aligned

Slight Rotation Creates Moiré

Figure 15.

Moiré example. The two patterns of lines should align exactly, as shown in the top figure. Rotating one creates a moiré pattern.

misaligned (Figure 15), and creates a pattern that is robust with respect to misalignment between separations. The resulting pattern is a mosaic of colors that includes each of the primaries (CMY) plus black (K) and all of the overprint combinations.

Originally made for mechanical printers by using special filters and films, these patterns are now digitally produced, as shown in Figure 16. In the figure, a 6 × 6 array of pixels is assigned to each halftone dot. This can create 37 gray levels, with 0 to 36 pixels turned on. To get 256 gray levels requires at minimum a 16 × 16 array, which is 2400 dpi for a 150 lpi screen. Creating rotated screens without introducing moiré patterns requires even more resolution. Therefore, traditional halftone patterns are only produced digitally as part of digital prepress.

Digital printers, which have much better alignment and often lower resolution, use other patterning techniques such as *dithering*. A dither pattern does not superimpose a regular grid like a halftone screen, but contains the same set of primary and overprint colors as a halftone pattern. Figure 17 shows a typical dither pattern from an inkjet printer. *Stochastic screening* is essentially dithering applied to offset printing technology.

When CMYK is specified digitally, the pixel values define the percentage of ink covered by the halftone dots, or *dot area,* expressed in percentages. Full coverage is 100%, a pattern that is half covered is 50%, and so on.

Because printing ink on paper is a mechanical process, simply imaging a 50% screen pattern on the printing plate, for example, does not guarantee a 50% printed pattern. In offset printing, this

Figure 16.

Digital halftone creation. Each halftone dot is created from many pixels.

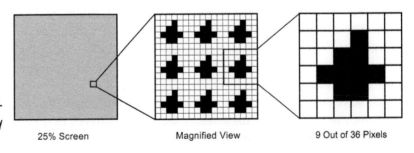

25% Screen          Magnified View          9 Out of 36 Pixels

difference is called *dot gain*, because the usual problem is that ink spreads as it is printed, making the dots bigger on the paper than on the printing plate. A typical offset printing press may have a 20% dot gain, which is a significant effect. In digital printers, dots can either be larger or smaller than the binary patterns used to create them, depending on the technology used. On a printing press, the halftones imaged on the plate have been preprocessed to compensate for dot gain: if a 50% screen is desired on a printer with 20% dot gain, for example, only a 42% screen is imaged. In digital printers, this correction is managed electronically with lookup tables.

It is easy to measure dot area with a densitometer. For a film that is a pattern of clear and opaque black dots, such as is used to image a printing plate, the dot area is exactly the inverse of the transmittance—a 60% halftone will transmit 40% of the light. For printed halftones on paper, internal scattering makes the correspondence much less exact. However, densitometry is still useful, especially for process control, where a way to measure the difference from a standard is all that is required.

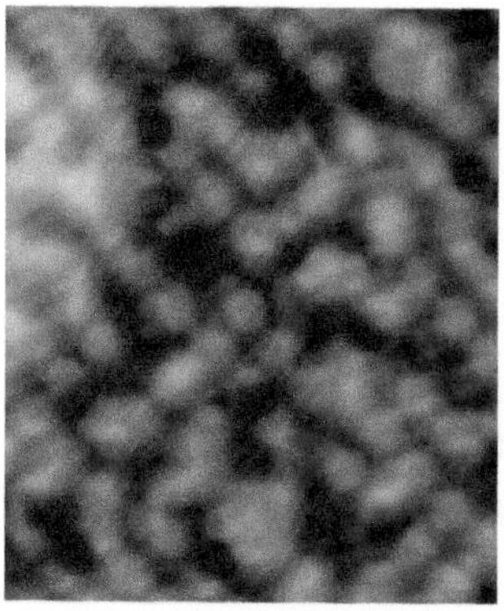

Figure 17.
A dither pattern for an inkjet printer.

# Subtractive Color Reproduction

Subtractive color reproduction means creating an image on a print or a slide that "looks like" the original. In traditional color reproduction, there was a tight coupling between image capture and reproduction, which made it possible to tune the capture filters to the output filters to give a predictable, if not ideal transformation. This is impractical for the star-shaped model common to digital color reproduction.

Subtractive color reproduction for digital color means taking an image encoded as RGB pixels and reproducing it on a subtractive color system. For printing, this means converting from RGB intensity values to CMY(K) values, usually specified as density or dot area, customized for a particular printing system. For a film recorder, it means converting from the input RGB to the RGB signals that will create the best color reproduction (which is ultimately CMY layers in film). In terms of device-independent color reproduction, it means mapping from the additive gamut of the input pixels to the subtractive gamut of the output system.

Subtractive color gamuts are very different from additive gamuts: they do not plot as nice triangles on the chromaticity diagram as is shown in Figure 18, which compares a monitor with a film and a print gamut. Even this figure understates the problem, as many colors that appear to fall inside of both gamuts do not. There can be a big difference in the brightness range obtainable on a subtractive gamut compared to an additive gamut, as the gamuts are quite different shapes in CIE XYZ tristimulus space, as shown in Chapter 9. As a result, sophisticated *gamut mapping* algorithms, which will be described further in the next chapter, are needed to transform from additive to subtractive gamuts.

Figure 18.

Comparison of a monitor, film, and print gamut.

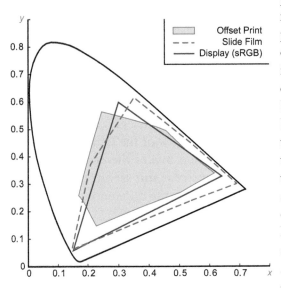

In additive color reproduction, both the pixels and the primaries function as tristimulus values. There is no correspondingly simple tristimulus model for subtractive color. The process of subtractive color depends on specific spectral distributions, including that of the ambient light, not just tristimulus values. In addition, subtractive color is typically dependent on a larger set of variables than additive color, including the light source, the reflectance properties of the paper substrate, exposure times, variations in film processing, ink consistency, the halftoning algorithms used, press settings ... the list can go on and on.

As a result, using colorimetric principles to characterize subtractive processes has not been a traditional part of the subtractive color reproduction industries, in contrast to video, which is based on additive color. The application of device-independent color to the graphic arts was hinted at in the late 70s, demonstrated in the early to late 80s, and is only now becoming common, driven by the star-shaped color reproduction process inherent to digital color.

The next sections discuss digital film recorders and printers which are the technologies of subtractive color, followed by a description of how such systems are typically characterized as part of device-independent color reproduction systems.

# Film Recorders

A film recorder writes an image on film, which is then processed to develop the picture. Film recorders are used to create slides and movies from digital imagery. They provide photographic prints as output for digital photo repair and retouching. In this latter application, a photograph is scanned, modified, then output to negative film so it can be printed just like any other photograph. Film recorders also provide film output for digital photography—many portrait studios now use digital cameras.

Film recorders combine a camera body with a small, high-precision grayscale monitor and a color wheel with red, green, and blue filters. The camera is mounted to focus on the display, and accepts regular film stock. The process of creating a slide first opens the shutter, images each of the red, green, and blue components of the image through the appropriate color, then closes the shutter and advances the film. Once the entire roll is exposed, the film is processed normally.

The monitors and optics used in film recorders are engineered to produce a perfectly rectangular image with sharp focus, even in the corners. Monitor resolutions are roughly 4000 pixels across for slide film recorders and 8000 across for 4 × 5 transparencies (though some slide recorders will image 8000 across also). The height depends on the ratio needed, which varies with the application. For example, the ratio for 35mm slides is 3:2. The intensity resolution for a good quality film recorder is 10–12 bits/pixel, rather than the 8 bits/pixel common for displays.

It takes a lot of engineering to create sharp images on film with a CRT. One common problem is blooming, where the pixel on the film is larger in size than desired due to overexposure. Another is *haloing*, which is most visible on negatives. In haloing, scattered light partially exposes the film around the pixel, creating an effect similar to film fog.

The earliest digital film recorders simply captured the video signal from a CRT, creating blurry, low-resolution images. Modern film recorders driven by a computer accept a wide variety of high-resolution image formats, and render Postscript or PDF as well. Their output is crisp and bright, matching the quality of traditional photography.

Some high-end film recorders use scanning lasers for exposure to create bright, sharp, high-resolution images. These systems are inherently slower, and much more expensive, than display-based recorders. Looking at film recorder products on the web, it is clear that speed is still a primary concern and selling point. No scanning system will ever be as fast as imaging a full frame at a time.

Film recorder output is processed using traditional, chemical photo processing. The resulting color depends on the film, the film-recorder settings, the processing, and ultimately, the color characteristics of the lighting or projection system. Therefore, it is very difficult to characterize a film output system unless strict control can be maintained throughout all the processing steps. *Sensitometery* is the field that creates metrics for control in photographic processing. For example, film can be precisely pre-exposed with light of known intensity to create a standard set of gray steps. Developing samples of this film along with the imaged film gives a standard by which to analyze the film development process.

Film recorders for slides used to be a staple of the computer graphics and digital imaging research professional, and were also used commercially for a wide variety of presentations. Now that digital projectors are cheap and bright, and digital video formats are becoming widely available, making slides with film recorders has been replaced by direct view of digital information for many presentation situations.

The use of digital matting, post-production, and special effects in cinema has created a growing market for movie film recorders. The digital color alternative is called *D-Cinema*, or *digital cinema*, which uses a very large, bright digital projector instead of projected film. At this writing, the quality of an all-digital solution for cinema, which involves both new cameras and new projectors, isn't as good as traditional film. Also, retooling all the multiplexes in the world with expensive new digital projectors doesn't have an economic model to support it. As a result, replacing film for "major motion pictures" won't happen soon. Other applications of moving images, from art cinema to corporate training, are quickly converting to the convenience of all-digital additive color by using high-quality digital video instead of film.

The use of film to create high-quality, photographic prints of digital, or digitally enhanced, images plays to the real strength of subtractive media—high quality color prints. The only competition is another subtractive technology, color printing, which rarely achieves the visual quality of photographic prints. The value of printing is in its speed, convenience, and ability to print on plain paper.

# Printers

Printing systems vary in size from small inkjet printers to room-sized gravure presses; they have in common the process of creating images by placing inks on paper. Figure 19 shows an image separated into its four *color separations*, one each for cyan, magenta, yellow, and black. These are printed one on top of the other to create a full-color picture.

Overlaying CMYK separations is common to all printing technologies. The interface between digital color and printing these separations is sufficiently different for commercial printing and for desktop digital color, however, that the two will be described separately. The digital interface to commercial printing, which takes place on printing presses, is the field called *digital prepress*. Digital printers, from personal desktop inkjet printers to shared xerographic sys-

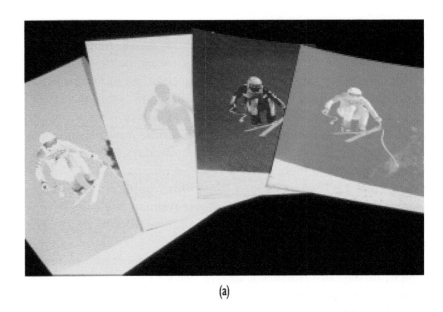

(a)

Figure 19.
Four color separations are overlaid to make a full-color image. (a) Prints of the cyan, magenta, yellow, and black separations; (b) Full-color image.

(b)

tems, are treated more as computer peripherals—digital output devices that are either connected directly to the user's workstation or accessed over a network.

Inevitably, the line between these two forms is blurring. There are digital printers as large and as fast as printing presses, and there are printing presses where digital signals are used to directly create the printing plates. For digital color reproduction, the distinction is primarily how the person creating the reproduction manages and manipulates the interface to the printing system. If you press a button on your computer screen and it prints, it's desktop. If you carefully package and ship off print-specific files, it's prepress.

## Commercial Printing

In commercial printing, a graphic designer creates the content for each printed page, consisting of colored images, illustrations, and text. This is transformed through the prepress process to the form needed to create printing plates. Twenty years ago, graphic design, prepress, and printing were practiced by different specialists. Digital technology has blurred these lines, especially by transforming prepress from a field based on film and chemical processing to a digital one. This section gives a brief introduction to commercial printing technology. Digital prepress is described in the following section.

There are three basic forms of printing presses: Letterpress, offset lithography, and gravure or intaglio printing. Each has a different type of printing plate and a different mechanism for transferring the ink to the paper.

In letterpress, the printing plate is raised where there should be ink, as shown in Figure 20, where black ink sits on the raised portions of the printing plate. The desired pattern is created by etching or cutting away all of the unwanted areas, as in woodblock printing. Ink is carefully applied to the remaining surfaces, then the plate is pressed directly onto the paper. This is how most people have been taught a printing press works, but letterpress is rarely used commercially. A variant of letterpress, called flexography, uses a flexible, molded

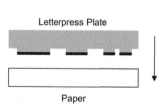

Figure 20.

Letterpress printing. Ink is applied to the raised surfaces on the plate, which is pressed directly onto the paper.

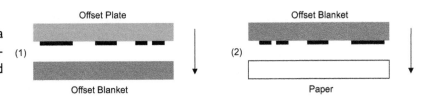

Figure 21.
Offset lithography. Ink adheres to a flat plate and is first (1) transferred to a rubber "blanket," and then (2) to the paper.

sheet rather than a metal plate, making it easier to press the ink onto the paper. It is often used for packaging materials.

In offset lithography, the printing plate is flat, thin, and flexible enough to wrap around a roller. The image is created by treating the surface of the plate, so that ink will selectively stick to it. The ink is then transferred from the plate to a rubber roller (or *blanket*), which is then pressed onto the paper. Figure 21 illustrates this process. In Step (1), ink is transferred from the flat plate to the offset blanket. In Step (2), the ink is transferred from the blanket to the paper. This two-step process, where the image moves first to the blanket and then to the paper, is why it is called *offset* lithography. Offset lithography is probably the most ubiquitous form of commercial printing. It is used for books (including this one) magazines, posters, and brochures. The art printing form called lithography uses similarly treated flat stones for printing, but the stone is pressed directly onto the paper. The rubber blanket both transfers ink more effectively, allowing offset printing on a wider range of papers, and allows the printing plate to be "right-reading" rather than mirrored, which helps reduce errors in plate production.

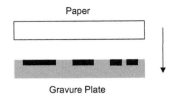

Figure 22.
Gravure or intaglio printing. Ink sits in depressions on an engraved plate or drum, and the paper is pressed down onto it.

In gravure printing, the ink is held in wells cut into the printing plate. Pressing the paper against the plate transfers the ink (Figure 22). Gravure, or intaglio, printing is derived from engraving, where patterns are carved into a metal plate. The printing plates of a commercial gravure printing press are cylinders, which are drilled in a pattern of dots to create halftone patterns. The deeper the hole, the more ink it will hold, and the darker and larger the dot when the ink is transferred to the paper. Gravure printing is used for long, high-speed print runs, such as catalogues and newspaper supplements.

Printing presses can print ink of any color, not just the standard *process colors*: cyan, magenta, yellow, and black. Process color is needed for full-color image reproduction, but text, logos, and illustrations may use special colored ink, called *spot colors*. If a printed document has no full-color images, a creative use of spot color can be more visually effective, as well as more cost-effective, than process color.

Black ink is always used in printing to expand the print gamut to improve the contrast. Recently, commercially viable techniques have been developed to use additional colored ink to expand the colorfulness of the gamut. One approach is to simply add a second printing of one or more of the yellow, magenta, and cyan inks, which improves the saturation of these colors. Duplicating magenta to improve saturated reds is common. Some systems print two similar colors of magenta or cyan. Pantone's Hexachrome and other extended printing systems add additional colors, such as orange and green, to stretch the gamut even further.

High-speed printing presses are mechanical wonders that move paper precisely through a series of inking stations, one for each color. The paper may be fed on a roll, called a *web*, or it may be fed a sheet at a time. When printing books or magazines, each press page contains several book pages, called a *signature*. A high-speed web press with eight printing stations can print four colors on one side of the page, flip the paper, and print four colors on the other side, then automatically fold the resulting signatures for binding, all while the paper is moving at speeds greater than 2000 feet/minute. Slower, smaller printers have fewer inking stations, and are usually sheet-fed. The smallest color offset printers print one or two colors at a time, and must be cleaned and refilled to print the additional colors.

# Digital Prepress

The goal of prepress is to take the original materials for a publication and create the image for each printing plate. The digital prepress process starts with the materials comprising a page: text, illustrations (line art), and images. These are composed into pages. Colored components must be separated and halftoned for the target printing system. Each set of pages is composed into a signature (a process called *imposition*). For each signature, as many color plates must be created as there are inks to be printed.

In offset printing, which is the technology readers of this book are most likely to encounter, the plates are imaged by exposing a photosensitive material coating their surface. Traditionally, they are exposed through high-contrast, black and clear film placed on the plate; more recent technological innovations allow direct, digital imaging on the plate itself.

An *imagesetter* uses a laser diode and special, large-format film to create a high-resolution binary image at the size of the printing plate. Typical imagesetter resolutions range from 1000 dpi (suitable for text and simple illustrations) to 2540 dpi, or sometimes even over 3000 dpi, for color halftone patterns. Input to the imagesetter is CMYK pixels; output is halftoned films suitable for plate making. The same imagesetter technology is used to write on thin, polyester plates for direct digital imaging. (Remember that in offset printing, the plate is thin, flexible, and flat.)

*Proof prints* are used throughout the prepress process to verify correctness and provide feedback to the prepress staff and the client. Traditional proofing processes work directly from the films created for plate making. Dupont Cromalin or Kodak Matchprint are examples of color proofing processes that use the plate-making films to mimic the printing process. Such processes give high color accuracy, but are expensive and slow.

Digital printers can be used to make proofs from the information sent to the imagesetter. The best color accuracy is achieved using high-end inkjet and dye-sublimation printers created for proofing, as their page sizes and color gamuts are designed to match commercial printing. Proofs created on desktop or networked printers can also be used, but require careful color management to give accurate color results.

*Soft proofing* is the term the printing industry uses for creating proofs on displays, which requires a CMYK to RGB color transformation. Even if this transformation were completely accurate, there are colors in printing that are out-of-gamut on displays, most noticeably golden yellows and forest greens. More subtle, but also significant, is that the appearance of a glowing display is different than a paper print. It tends to looks more vivid, especially if the lights are low enough for optimal viewing. For these reasons, soft proofing can never entirely replace printed proofs for high-quality color reproduction.

*Trapping* is the process of creating sufficient overlap between different color separations to avoid visible gaps due to misalignment on the press. Trapping is needed for illustrations, where colors may abut at a sharp edge, not for the continuously changing colors of an image. While there are many subtleties to the problem of trapping, the basic principle is to ensure there is always continuity across edges in at least one color separation. There are custom programs to automatically apply traps, and programs such as Adobe Illustrator include tools for manually creating traps.

# Digital Color Printers

Digital color printers are computer or networked peripherals that are either driven directly from imaging applications or sent as files in some page description language format. Desktop printers are usually inkjet; higher-speed networked printers are usually xerographic laser printers.

Inkjet printers spray tiny droplets of ink on a page through tiny nozzles in their *print heads*. The ink is enclosed in cartridges, which may have the print heads built right into them. The print heads step across the paper, which is moved underneath them to advance each new row. This is shown schematically in Figure 23. Each print head may have several hundred nozzles, and will print many dots at once.

There are two basic technologies for spraying ink droplets: thermal bubbles, called *bubble jet* (typically HP and Canon printers), and *piezoelectric* (Epson printers). In bubble jet printing, tiny resisters heat the ink, which creates a bubble. When this bubble pops, ink is sprayed on the paper, and fresh ink is pulled into the print head. In piezoelectric printing, piezoelectric crystals vibrate in response to

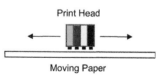

Print Head

Moving Paper

Figure 23.

Inkjet printer. Liquid ink is sprayed from a scanning print head onto a moving sheet of  paper. In this illustration, the paper moves perpendicular to the page.

electrical impulses, pushing the ink out of the nozzle. An actual printer dot may contain many droplets.

The quality of the reproduction on an inkjet printer depends strongly on the type of paper used. Contrast, saturation, and sharpness are all much, much better on bright white coated paper than on plain paper. Inkjet ink, which is quite fluid, soaks into plain paper, forming a thin, ragged spot. On glossy paper, it beads up, forming a round, thicker drop. Bright white paper will improve the appearance of any printing technology, as it increases the amount of light reflected back from the paper through the colored ink.

Xerographic printers, sometimes called *laser printers*, deposit layers of a plastic-like toner, which is then heated or *fused* to create a permanent print. The toner is applied, via electro-static charge, to a drum covered with photoelectric material called the *photoreceptor*, then transferred to the paper in the same pattern. Figure 24 shows the components of a typical laser printer. The image is created by using a laser to write on the photoreceptor, creating a pattern of charged regions. The toner is attracted by the charge, then deposited onto the paper in the correct pattern. After all four colors have been deposited, the paper passes through the fuser, which melts the toner.

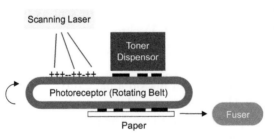

**Figure 24.**

Xerographic printer. An image is written with a scanning laser to deposit charge on a rotating photoreceptor. The toner sticks to the charged regions and is deposited onto the paper, which is then passed through the fuser.

Most laser printers have a single, drum-shaped photoreceptor surrounded by four color toner housings, and operate on one page at a time. The biggest and fastest printers, which can print 40–60 pages per minute, have a pipeline of pages passing through a sequence of toning stations, like a printing press.

While most digital printers use dithering or some form of halftoning to create variations in lightness, some have the ability to produce some grayscale on a dot-by-dot basis. For example, an inkjet printer may create each spot with multiple drops. Xerographic printers may combine a small amount of grayscale with spatial patterns. Dye-sublimation photo printers evaporate dyes in varying densities onto photographic paper, giving a photographic-like grayscale.

# Characterizing Subtractive Systems

Characterizing a subtractive color system means creating a transformation between some input pixel values and a visual color space such as CIE XYZ or CIELAB. This can be used to create the transformation between characterized RGB pixel values and the output colors. As emphasized throughout this chapter, there is no simple, mathematical model that links subtractive color to colorimetry as there is for additive color.

While specialists in specific technologies can create models that predict subtractive color from colorimetric measurements, the process of creating a device-independent color specification for digital color reproduction is usually empirical. First, the subtractive process is encapsulated at some point where there is a clear set of digital inputs. Then, a set of test samples is produced by systematically varying the input values to span the color gamut of the system. These samples are measured to produce a 3D table of CIE tristimulus values (or more commonly, CIELAB), which can be interpolated for intermediate colors. This gives a mapping from the input values (typically RGB or CMY) to a device-independent color space, and also defines the output gamut. Numerically inverting the table creates the inverse mapping.

Figure 25 illustrates this for a digital desktop printer. Such printers can logically be driven with RGB inputs, which are converted to CMYK for the printer by some combination of the device driver

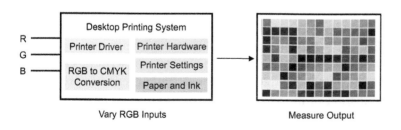

Vary RGB Inputs         Measure Output

Figure 25.

Illustration of subtractive characterization for a desktop printer. Varying the RGB inputs samples the gamut. The results are measured to created the characterization.

and the firmware in the printer. A test pattern is created by varying the input RGB to create a set of printed patches. These are measured to produce the RGB to CIE XYZ translation table, which serves to characterize the complete printing process to a device-independent color specification. Note that included in the characterization are the physical characteristics of the printer, the paper, the printer's settings, and the driver software. Changing any of these can change the color and invalidate the characterization: strict process control is critical to maintain accuracy.

Printing systems that accept page description languages such as Postscript (Adobe) or PDL (Hewlett-Packard) must correctly reproduce color specified in a variety of forms. Therefore, the printer characterization may be defined closer to the printer hardware so it can be combined with other characterization data to create all the needed color transformations. Such a characterization may be made for a specific black-generation and gray-balancing algorithm, in which case the input is CMY. Or, black may be allowed to vary like the other primaries, creating a larger, but more general, 4D table. In this case, the input is CMYK.

Characterization for film recorders almost always includes the film processing, which introduces a large, difficult to control variable in the color reproduction process. While sensitometry makes it possible to detect variations, there is no way to "redevelop" the processed film. Processed negatives, however, can be printed and reprinted to achieve consistent positive images—this applies both to paper prints and to movie film, which is shot as a negative and printed to create positive film. There are, of course, economic constraints on reprinting in real production systems.

Projected film would be most accurately characterized by projecting it and then measuring the color through an instrument with a lens. This is rarely practical. More often, the film is backlit and measured with an instrument pressed against it.

Because subtractive color is often quite non-linear, it is common to measure 1000 patches or more to create an accurate characterization. Sometimes, subtractive characterizations can be derived from a combination of models and data to reduce the number of measurements needed.

The use of large tables in color management systems has created a market for fast, automated color measurement equipment designed to measure a large number of color patches printed on paper. These systems either use an x-y positioning stage to select each patch, or scan the patches printed on a strip of paper. These are designed primarily for the professional graphic arts industry, and cost several thousand dollars for the hardware, and another several thousand for the software. Similarly, there has been extensive work on high-speed interpolation algorithms and hardware, primarily for digital printing applications.

## Summary

Twenty years ago, digital printers were rare, expensive, and only worked well when the local specialist was near by. Today, a $150 inkjet printer gives remarkable quality, especially for a system that is designed to work "straight out of the box." On high-quality, glossy paper, it can produce bright, vibrant color pictures, albeit very slowly. Mid-range xerographic printers have made color reports, newsletters and other office documents the standard in many workplaces. Digital prepress has simplified the interface to traditional printing. Once strictly a professional craft, printing is now practiced by anyone with a computer.

Photography has long been part of the consumer market, from vacation snapshots to the dedicated amateur artist. Digital photography, which breaks the tight coupling between image capture and color reproduction embodied in a film camera, is converting photographers to digital color specialists. In the consumer market, creating prints from digital photos more often involves a printer than a film recorder. Film recorders, however, are evolving to meet the increased use of digital imaging in cinema, portrait photography, and photo restoration. And, desktop film recorders for creating prints from digital photographs may yet create a consumer market for digitally-driven film output.

Subtractive color systems, print, and film render color by filtering white light through layers of inks or dyes. In print, the filters are the printing inks, which are deposited on white paper. In film, the colors are dyes deposited in layers in the film stock. Light is passed through these filters either by direct transmission (as in transparency film) or by reflecting it off the white background of a print. Film systems vary dye density to encode grayscale, whereas printing systems typically use fine patterns of colored dots, either halftone or dither patterns.

Cyan, magenta, and yellow are the "primary colors" of subtractive color systems. Ideal primaries would selectively modulate independent parts of the input spectrum. Actual colored inks and dyes are not independent filters, and the resulting cross-coupling can make even creating a uniformly neutral ramp from black to white a challenge.

There is no simple model for characterizing subtractive color. This, plus the large number of process-specific variables inherent in subtractive color reproduction, forces a purely empirical approach to characterization: encapsulate the process, sample the input, measure the output. Maintaining such a characterization demands strict process control.

Commercial systems for creating profiles for subtractive systems are more expensive than those for displays. Fortunately, the default profiles for desktop printers are often very good. However, there are services that will create custom profiles for those who need a more accurate characterization. The service supplies a digital test pattern to the client, who prints it using their preferred paper, ink, and settings. The resulting print is returned to the service, who measures it to create a custom profile, which can be used with any standard color management system.

Understanding subtractive color can provide insight into the process of creating high-quality images on print or film. But, as was stated at the start of this chapter, understanding subtractive color doesn't address the key concern of most users of subtractive color systems: making good-looking prints from RGB pixels. This, and more broadly, the problem of managing color across different digital media, is the topic of the next chapter.

# 9
# Color Management Systems

Color management systems combine the metrics created using device-independent color representations with the color reproduction principles described in Chapter 5. First hinted at in the late 1970s and then demonstrated in the 1980s, device-independent color management is now an established commercial domain, with support in both Microsoft and Apple operating systems. This is the last of the five color reproduction chapters. It provides a description of color management systems, the principles behind them, and some examples of their application.

## Introduction

Color management systems were designed for the star-shaped reproduction process introduced in Chapter 5, which is shown again in Figure 1. Colors and images from various sources can be captured, manipulated, and reproduced in a variety of forms. To have a reliable, controllable color-reproduction process, it is critical to under-

Figure 1.

The star-shaped digital color reproduction process, with all color transformations marked with a red box.

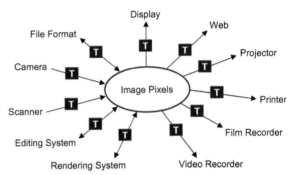

stand and manage all of the color space transformations that affect the reproduction, indicated by the red "transformation" boxes in the figure.

Color management systems use device-independent color representations to characterize devices and images to a visual standard such as such as CIE XYZ or CIELAB. This characterization is used to control the color reproduction process, providing an algorithmic way to describe any path through the star.

Color management systems are not easy to implement or maintain. Accurately characterizing all colors and devices is the first challenge, as it requires transporting the technology of color measurement into the workflow of color reproduction. Even assuming accurate characterizations, there remains the problem that different media have different color gamuts—the desired color may simply not exist on the target device.

Color management systems are based on the principles of color measurement, but are trying to solve a problem that is ultimately color appearance mixed with the craft and economic constraints of color reproduction systems. While they are a good first step at introducing scientific and engineering principles to the image reproduction problem, they are inevitably an imperfect solution, and their technical complexity can be intimidating. Some of this complexity is inherent, but far more comes from the problem of retrofitting color management into existing applications and work processes. The result is multiple, and often conflicting, ways to specify and control the digital color reproduction process.

Giorgianni and Madden have written a book called *Digital Color Management*, which is an excellent in-depth reference for this topic, particularly for implementers of color management systems. A more recent pair of books, *Colour Engineering: Achieving Device Inde-*

*pendent Colour* and *Colour Image Science: Exploiting Digital Media,* are edited collections of up-to-date articles on the technical aspects of device-independent color.

A detailed description of color management aimed at the graphic arts market is *The GATF Practical Guide to Color Management,* by Adams and Weisberg. (GATF stands for Graphic Arts Technical Foundation.) This book, published in 2000, includes specific references to color management software and its implementation in desktop systems and applications. The color workflow and management parts of Phil Green's book, *Understanding Digital Color,* also provide an up-to-date, practical guide for professional users in the graphic arts. The explanation in *Adobe Photoshop 6.0 for Photographers* provides a clear explanation of color management in terms of RGB color spaces, though not at a highly technical level. In contrast, Dan Margulis, in his books and teachings on Photoshop and color correction, blasts device-independent color management as inadequate for truly professional control of color in printing—color management is still not universally appreciated.

This chapter begins with an overview of color management systems and their components. It then discusses how to apply these principles to color reproduction systems.

## Overview of Color Management Systems

The color management problem is often stated as: Given a set of digital input and output devices such as printers, scanners, monitors and film recorders, how do we convert colors between them without creating a customized transform between each pair of devices? The answer is to use a single, device-independent color space (such as those defined by the CIE) as the common interchange format. For each device, create a set of data called a *profile* that maps between its native color space and the interchange space, or Profile Connection Space (PCS). Figure 2 deliberately mimics Figure 1. But in this case, all devices point into and out of the PCS, connected through their profiles (blue boxes). This is the canonical model for

Figure 2.
The model for a color management system, where all components are linked through profiles and the profile connection space.

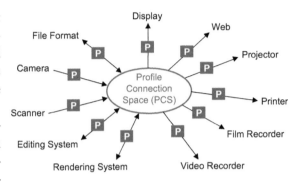

Figure 2.
The model for a color management system, where all components are linked through profiles and the profile connection space.

color management systems: characterized devices, linked via their profiles through the profile connection space.

Given an image displayed on one device, it is reproduced on another by mapping it through the PCS as shown in Figure 3(a). If all images and devices had the same gamut, this would be sufficient. But, as has been stated many times before, different devices have different gamuts. A simple conversion into and out of the PCS can create colors that are out-of-gamut on the target device. More subtle, but equally important, different devices have different appearance characteristics—most critically, different white and black points. A simple transformation will preserve these differences, rather than mapping black to black and white to white, which is important for good image reproduction. *Gamut mapping,* as shown in Figure 3(b), adds a transformation that maps one gamut into another.

The gamut of each device can be defined as a volume in the PCS. Similarly, images have gamuts, usually described as a cloud of points in the PCS. There are two forms of gamut mapping that can be used in image reproduction; the most common maps one device to another and uses this mapping for all images, which are assumed to be defined with respect to the source gamut. A less common form maps the image gamut is mapped to the target in a manner that is unique for each image. In the first form, a spe-

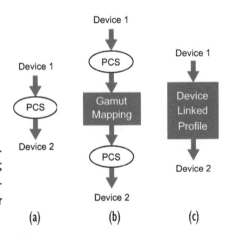

Figure 3.
The process of managed color. (a) Transformation through the PCS; (b) transformation plus gamut mapping; (c) combined transformation, or device-linked profile.

cific color in a source gamut will always map to the same color on the destination gamut. As a result, transformations into and out of the PCS can be combined with the gamut mapping transformation and represented as a single profile, sometimes called a *device-linked profile*, as shown in Figure 3(c). In the image-specific form, each transformation step needs to be performed separately.

A color management system consists of a mechanism for managing profiles, plus a *Color Management Module* (CMM) to perform transformations, including gamut mapping, using the profile information. As in all color reproduction systems, there is no unique "correct" way to define these transformations; the best mapping will be both image and application specific. As a result, different color management modules will give different results, even using the same profile information.

All color management systems include ways for developers and users to influence their operation; the most basic is selecting the right profile information for the specific devices used. Additional controls are available to influence the gamut mapping, which is discussed later in more detail. Unfortunately, determining how to reliably set up and control a color management system can challenge even dedicated professionals. Even within Apple's ColorSync, a decade-long effort to provide a simple, unified model for color management in the Macintosh operating system, there are multiple, competing controls. For example, to print a displayed image from an Adobe application, the user can use the profiles and transformations defined in the ColorSync system user interface, use those provided by the printer driver, or use those provided by Adobe. All use color management principles, but they give different results.

## Profiles

A profile is a data file that contains characterization data, as has been described in previous chapters for additive, subtractive, and image capture systems. Profile information is most often in the form of 3D lookup tables that map device-specific colors to CIELAB or CIE XYZ.

Trilinear or tetrahedral interpolation is used for intermediate colors. For displays, an alternative representation is a $3 \times 3$ matrix plus three intensity transfer functions (stored as tables), as described in Chapters 3 and 7. The profile may also contain additional data such as the white point, the black point, and other information useful for applying or identifying the profile.

The International Color Consortium (ICC) was formed to support the exchange of profile information across systems. This industry consortium has defined a standard profile format that is now commonly used by all commercial systems (www.color.org). Without such a standard, different color management systems would require different data files.

Image colors must be specified with respect to the PCS. If the profile is included in the image file, it is called an *embedded profile*. Or, the image can include only the name of a profile, whose full specification is stored in the color management system. Alternatively, image formats such as Flashpix or MPEG include a color space (equivalent to a profile) as part of their specification.

In theory, an image can be specified independently of any specific device, forming its own gamut in the PCS. In practice, images are usually tagged with the profile of a scanner or display, depending on how they were generated. While primarily pragmatic, such a specification has some technical advantages over a more general one. A device profile defines a full color gamut, including the neutral axis. Having the full gamut also makes it possible to determine how saturated a color is relative to the gamut's extent. This can provide a better basis for gamut mapping than just the image gamut. It also ensures that all images from the same display or scanner will be transformed in a similar manner—an advantage for some applications.

A particular color device can have many profiles associated with it. While most desktop printers have only one type of ink, they accept different papers, which changes the colors produced. Displays have different white points and different transfer curves. Scanner profiles are generally accurate only for one set of colorants, such as a specific photographic film. It is the responsibility of users to select the profile that best matches their specific configuration.

Profiles capture device colors for a single set of viewing conditions, the most critical component of which is the ambient illumination. This is most significant for subtractive color prints, as the color is a direct function of the ambient light. For best results, the profiling viewing conditions should match the application viewing conditions as closely as possible.

In practice, there is almost always a mismatch between the profiling and the final viewing conditions. Most color instrumentation for profiling subtractive media comes with its own embedded light source, and is designed to eliminate all other light. Instruments for profiling displays are usually designed to be pressed against the screen, eliminating all ambient light. This makes it easy to create a consistent profile that only varies with the state of the device. However, it does not match any natural viewing conditions, which have a more complex blend of lighting and other factors that affect image appearance.

To what degree controlling the viewing conditions matters depends on the image reproduction goals. At one extreme, if the goal is simply to match trichromatic measurements, any variation in viewing conditions would be significant. More often, the transformations introduced by gamut mapping are sufficiently large that variations in viewing conditions are almost insignificant. For color-critical applications, it is important to create a controlled viewing environment for all critical evaluation. While this environment may not match the profiling environment exactly, it provides an important consistency for judging the quality of the color reproduction.

## Gamuts and Gamut Mapping

Different devices have different gamuts. The most striking differences are between those of additive systems, such as displays or projectors, and the gamuts for subtractive systems such as print or film. This can be clearly seen in Figure 4, which shows a printer and a monitor gamut plotted on a chromaticity diagram. As expected, the monitor gamut is a triangle, but the printer gamut is quite irregularly shaped. In this example, neither gamut completely encloses the other.

A monitor (sRGB) and a printer (Matchprint proofing for offset printing) gamut plotted on a chromaticity diagram. The white points are marked: D65 for the monitor, D50 for the printer.

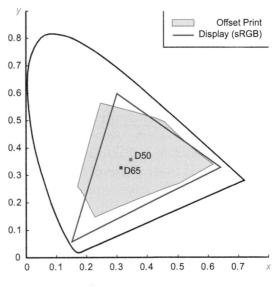

A true comparison of gamuts can only take place in three dimensions, a point often lost in the convenient use of the chromaticity diagram. Figure 5 shows the same gamuts as Figure 4, but plotted in three dimensions as xyY. This shows the gamuts rising above the surface of the chromaticity diagram, in proportion to their luminance. The wire frame is the monitor, the solid is the printer. The monitor gamut clearly has high-luminance colors that project to the same points on the chromaticity diagram as colors in the print gamut. That is, in 2D, they appear to be inside both gam-

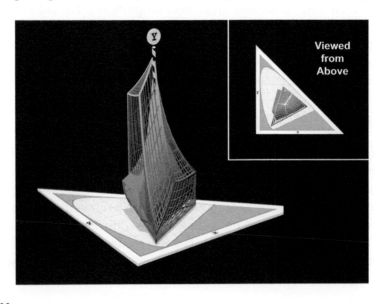

The gamuts from Figure 4 plotted in xyY to show the difference in luminance values as well as the chromaticity. Viewed from above (small inset), it reduces to the chromaticity diagram shown in Figure 4. (Images by Bruce Lindbloom, www.brucelindbloom.com).

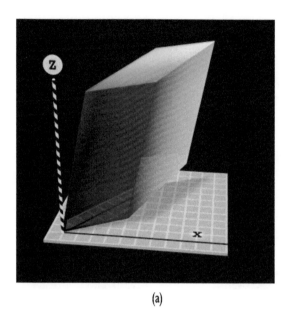

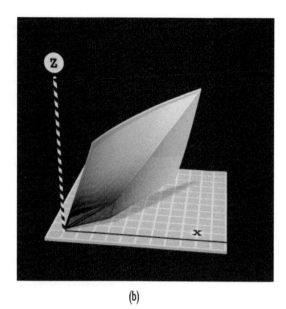

(a)                                          (b)

Figure 6.
The sRGB (a) and Matchprint (b) gamuts plotted in XYZ space. (Images by Bruce Lindbloom.)

uts, but in three dimensions, they are not. The small inset figure shows the 3D model viewed from above, revealing the same chromaticity diagram as in Figure 4.

Figure 6 shows these same two gamuts plotted in CIE XYZ tristimulus space. The gamut for the monitor is a regular shape reflecting the linear transformation between RGB and CIE XYZ for additive color systems, as described in Chapter 3. The print gamut, however, is distinctly non-linear; while there is some hint of the original color cube, each face is a different size and some of the edges are distinctly curved. The gamut is actually concave as well as bent.

In color management applications, gamut data are usually converted to CIELAB, which normalizes the gamut to a common white (but not necessarily black) value. This conversion also creates a perceptual organization of the color data. Figure 7 shows a monitor and a printer gamut in CIELAB space. Gone is the simple geometric relationship to the RGB color cube—both gamuts are highly nonlinear. The vertical (L*) axis is lightness; saturation increases radially, and hue is mapped around.

201

The sRGB and MatchPrint gamuts, plotted in CIELAB space. (Image by Bruce Lindbloom.)

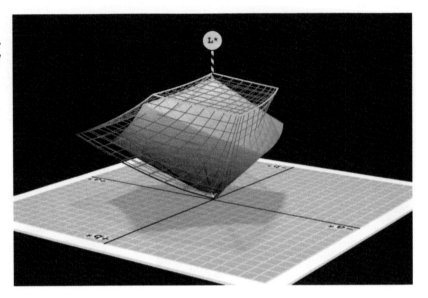

CIELAB must be computed with respect to a reference white (see Chapter 2). For displays, it is common to use the display white point, which means that white will map to L*, a*, b* = (100, 0, 0). In Figure 7, the reference white for the print gamut was measured from a highly reflective white surface designed to reflect all of the ambient illumination. This was done so that the paper color could be included as part of the characterization. As a result, the print gamut white does not match that of the monitor, nor does the black. These differences will have to be accommodated in the gamut mapping. As in Figure 5, is easy to see in this figure the brightness differences between the two gamuts, and the fact that neither gamut wholly contains the other.

Figure 8 shows an image gamut projected as a scatter plot on the a*b* plane, along with the outline of a monitor and a print gamut. It is easy to see that this image, which was designed on a display, contains many out-of-gamut colors with respect to the printer.

Most commercial color management systems perform gamut mapping in CIELAB space. After the neutral axes are aligned, out-of-gamut colors are projected toward the center of the gamut in a way

that reduces satura-
tion, and to a lesser
extent, brightness,
while maintaining
hue. While in theory
the mapping could be
from the image to the
target gamut, in prac-
tice, the mapping is
defined between de-
vices. This is not sim-
ply for computational
convenience—it also
provides some of the
constraints needed for

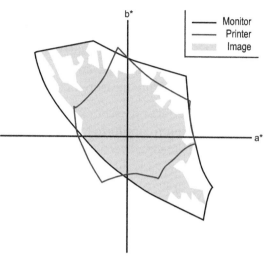

Figure 8.
The gamut of a colorful image designed on a display, plotted on the
a*b* plane, along with a monitor
and a printer gamut.

making a good reproduction. For example, the neutral axis is well-
defined for a device, but not necessarily for an image.

No single gamut-mapping algorithm is adequate for all images.
ICC color management defines a limited set of controls for gamut
mapping, leaving the user to adjust the original image (or the device-specific resulting image) to achieve the final aesthetic goal. These
are formally defined as *rendering intents* that provide four general
categories of gamut mapping—all except absolute colorimetric align
the neutral axes of the two gamuts, mapping white to white. Whether
to also map black to black is often a separate option.

Perceptual: Create an aesthetic mapping for images. Maps all
colors smoothly into the target gamut, though those outside
of the gamut will move relatively more than those inside.

Saturated: For charts and illustrative graphics that contain solid
saturated colors. Maintains the relative saturation values of
colors, and may sacrifice smoothness to maintain saturation.

Relative Colorimetric: Minimal color transformation. Usually
projects the out-of-gamut colors to the gamut surface with
little or no mapping of in-gamut colors. The default mapping for older printing systems.

203

**Absolute Colorimetric:** A pure colorimetric mapping. Matches measured value to measured value, without even aligning the neutral axes (applies to profiles specified in CIE XYZ). This is suitable only for logo colors and other specialized cases where the absolute measured color value must be preserved.

Not all interfaces present exactly these choices, but there is usually some option to optimize for photographic versus more illustrative images (perceptual versus saturated). These choices may be only presented in the "advanced" section of the controls for the color management system.

Research in gamut mapping explores better ways to match the appearance of images across media. Current algorithms are far from ideal. Mapping color designed on a blue-white CRT display to accurately match colors perceived on a print illuminated by a table lamp is still beyond the scope of current color management systems—the white point transformation in CIELAB introduces unwanted color shifts. Future systems may include the use of improved color appearance models such as CIECAM02, or those that evolve from it.

Even the best color appearance models, however, cannot generate a satisfactory match for severely out-of-gamut colors. It is an inevitable part of successful cross-media design to understand and accommodate gamut limitations in the design itself.

# User's View of Color Management

The color management system implementer thinks in terms of device profiles, profile connection spaces such as CIELAB, and transformations between color gamuts. Most digital color users, however, want to think in terms of a common *working space*, which is usually some form of RGB color space, plus transformations into and out of this space, as shown in Figure 9. Images or digital colors are created and manipulated in the working space. Colors and images are imported into this space, either using cameras and scanners, or algorithmically, as in computer graphics. Displays and other output devices are used to view this space. The transformations important to the user are those that convert colors and images into and out of the working space. These include not only the device profiles,

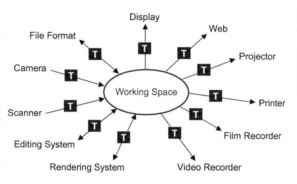

Display
Web
File Format
Projector
Camera
Printer
Scanner
Film Recorder
Editing System
Working Space
Rendering System
Video Recorder

Figure 9.
The user's view of color management, where all transformations are defined with respect to a common RGB working space.

but any settings that affect the transformations using these profiles. These combined transformations are represented as purple boxes in the figure.

The choice of working space depends on the application, both in terms of constraints on the RGB space (size, linearity), and whether the images and colors created in this space need to be exported, or merely displayed and printed. Once the space is chosen, all transformations in and out of the space need to be carefully defined and controlled. The next sections describe the considerations for picking an RGB working space, and how to define and control the color transformations.

# Picking an RGB Working Space

To review the principles first described in Chapter 3, an RGB color space can be defined by the color and magnitude of its primaries, typically expressed as the chromaticity of its primaries plus the white point and magnitude of white. Pixel values can either describe linear or nonlinear values with respect to intensity or luminance.

The ideal primaries for the working space depend on the application. It is most convenient to make these match the user's display, as this makes the entire working space visible without gamut mapping. However, in the graphic arts, where the work flow often progresses from scanned photographs to prints, it is better to choose a larger RGB space that completely contains both the print and the display gamuts.

Whether a working space should be linear or non-linear also depends on the application. Rendering and image processing algorithms should always be performed in linear intensity spaces. However, the vast majority of users work in a non-linear space defined by their display's characteristic transfer function. While linear spaces are required for many computations, non-linear spaces are more perceptually uniform for direct manipulation of colors and images.

If a display-oriented space is sufficient, one good solution for a working space is to use sRGB, which was described in detail in Chapter 3. Many programs and printers include sRGB as an option for specifying RGB colors, eliminating the need for embedded profiles. The sRGB color space is similar to the digital video color space (ITU-R BT.709) and uses the same primaries, though the specification of the non-linear transfer curve is slightly different.

It is easy to configure a CRT display to match sRGB, which makes the display and the working space identical. Typical CRT phosphors already match the sRGB primaries, as shown in Chapter 7. Therefore, setting the white point to D65 and the transfer curve to be gamma = 2.2 is sufficient to create an sRGB display. Both the white point and the gamma curve should be measured for accuracy rather than depending on values defined by the display controls, especially if the display is more than a year or two old.

Most PC display systems assume a gamma = 2.2 curve, making them a good match to sRGB. On a Macintosh, however, the display system modifies the effective display gamma to be 1.8. While it possible to set a Macintosh system to use a gamma of 2.2, it is contrary to standard practice and makes displayed colors and images designed for the Macintosh appear too dark and saturated. This makes it is less convenient to use sRGB as both the working space and as the display space on a Macintosh. However, the conversion between two gamma curves is straightforward, and many graphic arts systems support both "gamma 1.8" and "gamma 2.2" displays.

Flat panel displays, especially laptop displays, are not sRGB compatible because they have different primary colors. However, it is straightforward to transform these display colors to sRGB because their gamuts fall inside the sRGB gamut.

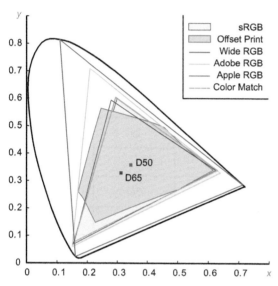

Figure 10 shows a collection of RGB working spaces plotted on the chromaticity diagram along with a print gamut. It is easy to see the print colors that are out-of-gamut with respect to the display-oriented sRGB, as well as the display colors that are out-of-gamut with respect to the print, as was also illustrated in Figures 4–7. The "WideRGB" gamut, whose primaries lie on the spectrum locus, easily encompasses them both, but includes many colors that are out-of-gamut on both devices. A better compromise is shown by Adobe RGB, which appears to include both. However, it is impossible to completely evaluate whether one gamut is fully contained in another by looking at only two dimensions, as was shown in Figure 5. The values for Figure 10 are shown in Table 1.

In summary, the most convenient working space to use is by far the one that matches the user's display, as this makes the entire working space visible. It also makes the transformation from working space to display easy and fast, which is important for interactive applications. If a linear space is needed for computational reasons, it is simplest to derive it from the display's RGB space, so that only the intensity transfer function varies between them. If, on the other hand, there is a need to optimize for print or film, then it's important to chose a larger RGB space to accommodate the colors missing on the display.

Figure 10.

A collection of different RGB working spaces commonly used in the graphic arts, along with the MatchPrint gamut. The associated values for the RGB spaces are shown in Table 1.

| Color Space | | Red | Green | Blue | White | Gamma |
|---|---|---|---|---|---|---|
| sRGB | x | 0.640 | 0.300 | 0.150 | D65 | 2.2 |
| Windows Default | y | 0.330 | 0.600 | 0.060 | | |
| Wide Gamut RGB | x | 0.735 | 0.115 | 0.157 | D50 | 2.2 |
| (700/525/450 nm) | y | 0.265 | 0.826 | 0.018 | | |
| Adobe RGB (1988) | x | 0.640 | 0.210 | 0.150 | D65 | 2.2 |
| | y | 0.330 | 0.710 | 0.060 | | |
| Apple RGB | x | 0.625 | 0.280 | 0.155 | D65 | 1.8 |
| Macintosh Default | y | 0.340 | 0.595 | 0.070 | | |
| ColorMatch RGB | x | 0.630 | 0.295 | 0.155 | D50 | 1.8 |
| (P22-EBU) | y | 0.340 | 0.605 | 0.077 | | |

Table 1.

The data values for the RGB color spaces in Figure 10. The x, y chromaticity values for D65 = (0.313, 0.329); D50 = (0.346, 0.369).

The sRGB color space has the advantage of being a display-oriented space that is also recognized as a de facto standard for color images in the desktop and web environments. It is a good working space for desktop color users, many of whom will be able to make their displays match sRGB as well. Macintosh users and those using flat panel displays, however, cannot easily match their displays to sRGB, though they can still use it as the working space.

## Specifying Color Device Transformations

The key to reliable color reproduction is to create a fixed transformation for each device and maintain it. There may be a profile and a CMM involved, but they are only part of the process. Anything that affects the colors, from the hardware drivers to the light shining on the screen, are part of the color device transformation.

208

Let's look at the components that can affect a color transformation, which are shown schematically in Figure 11. At the bottom level is the physical hardware, such as the printer, scanner, or display. All color reproduction hardware is subject to variation over time. *Calibration* is the process of ensuring the hardware is working within spec, and may involve returning it to a standard configuration, such as a specific white point and transfer function for a display.

Most color hardware will operate in a number of different modes. Displays have brightness or contrast controls. Cameras have settings that define the white-balancing algorithms used, and whether or not the flash is enabled. Depending on their size, printers may have an entire front panel full of controls for contrast, color balance, and more. The physical materials such as ink, paper, and film are also part of the setup for printers and film recorders. These *hardware settings* must be set and protected from change—putting tape over display controls is traditional in color critical environments.

Digital color hardware interfaces to the operating system through *drivers*, and these drivers include *driver settings* that affect the color. Call up the control panel for a display, or a printer, and marvel at all the options—all of these must be stabilized for stable color. If the driver provides an option to save the current settings in a file, using this option will help ensure consistency. The settings that affect the intensity transfer function, or gamma curve, for a display are particularly important. These settings operate by changing tables in the display controller hardware. Applications, including display profiling software, can also change these tables, causing a global change in display appearance.

*Applications*, especially those developed before OS support for color management, may include their own explicit and implicit transformations for color. Image editing and viewing programs may offer options that change the display of the color, but only inside of the application. As a result, the application is part of the color transformation as well.

*Viewing conditions*, which are defined primarily by the lighting, affect color appearance for all media. The adaptability of human

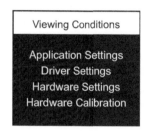

Figure 11.
The layers of components that affect color device transformations.

vision keeps us from being terribly aware of even dramatic shifts in lighting for casual viewing. Critical viewing, however, must have a controlled environment, and even casual viewing should have a consistent one. Displays should be viewed in a dim room, without light shining directly on them. Prints should be viewed with standard, uniform lighting. A graphic arts viewing booth, (Figure 12), provides both a standard light and a neutral gray surrounding environment for viewing prints, but even using a desk lamp on a table provides some consistency. Projected color, whether slides or digital, depends

Figure 12.

A graphic arts viewing booth, with a selection of light sources. The walls are neutral gray. (Image courtesy of Gardco.)

on the color of the bulb (which will change with age) as well as any light falling on the projected image. As in print viewing, the degree of control can vary with the nature of the application, but consistency is always important.

In summary, changing any part of the color transformation will potentially change the color. It can't be said too often—control and consistency are key to successful color reproduction.

# File Format Transformations

The color space underlying an image or video file can either be embedded in the file, or implied by the format. While there has been an on-going effort to establish standards for embedding a precise RGB specification in image files, this has had little practical effect as of yet. For image formats, only programs designed for the graphic arts (such as Quark Express or Adobe Photoshop) routinely create and read image file formats with embedded profile information. More commonly, the image format is defined relative to a standard RGB color space. This is the approach in video and in formats like the EXIF image format from JEIDA (Japan Electronic Industry Development Association), which combines JPEG encoding with an sRGB color space specification for digital photography.

Whether the profile is embedded or implied, the principles described in Chapter 3 can be used to convert from one RGB space to another. An additional consideration when transforming files is whether the conversion will result in colors that are out-of-gamut for that space, which must be clipped, or more ideally, gamut mapped into the destination space. When sending colors to an output device, there is no choice but to approximate such colors. Within the context of the working space, there is the option to try to maintain all of the original information, as described further in the discussion of Photoshop color management in the next section.

Many digital images have no color space information either included or implied. In this case, some intelligent assumptions must be made. The color space for images created by video cameras is non-linear, with phosphors defined by ITU-RBT.709. There are no similar standards for digital still cameras, but it is common to encode brightness non-linearly. Computer graphics formats are one of the few image formats that assume a linear color space, as described in Chapter 10. As described in Chapter 3, any image or individual colors created on a typical desktop display should be treated as non-linearly encoded. The sRGB encoding is a good default for images created on a CRT driven by Microsoft Windows or Linux, but Apple's operating systems use a gamma = 1.8 rather than a gamma = 2.2 transfer function by default. Colors defined on a flat panel are also non-linearly encoded, but have different primaries than a CRT display. Laptop colors are significantly less saturated than those of desktop displays.

# Color Management in Photoshop

Adobe Photoshop is one of the most commonly used tools to prepare images for printing. It is also a powerful image editing tool. This combination forces its color management model to consider both difficult cross-media transformations (display to print), but also the problem of interactively editing managed color. As a result, it provides a good example of the issues that must be addressed in color

211

management. Photoshop color management and its user interface have undergone many revisions since color management was first introduced in Photoshop 4.0. The following are my own observations based on Photoshop 6.0, and have not been validated or certified in any way by Adobe.

Figure 13 is a screen snap of the Color Settings manager. The specific values displayed are not a coherent set, but are chosen to illustrate the different options. This section will simply step through and explain the options, and what purpose they serve.

The first field, called "Settings" is a name assigned to this collection of settings, in this case "Illustration Settings." Settings can be saved and restored via files, using the Load and Save buttons. Photoshop comes with several default settings files, including: European, Japanese, and U.S. Prepress; web graphics; and backward compatibility settings for previous versions of Photoshop. Previous sentence is mis-punctuated in the page proof. This should be more clear. Managing settings in a file like this is an excellent way to ensure they remain consistent.

The next set of fields is called "Working Spaces." While this chapter has emphasized a single, RGB working space, Photoshop defines one space for each of RGB, CMYK, Gray, and Spot colors. This better serves the needs of those members of the graphic arts community that want or need to work in CMYK and spot color ink percentages directly. These settings define the default working spaces for new images. They also provide the specification for conversions between color spaces, such as RGB to CMYK, another important function of Photoshop.

The RGB working space is exactly as described in this chapter. In this illustration, it is set to sRGB, but it can also be set to any of a long list of predefined settings or set by loading a custom ICC profile. The option "Custom RGB" opens a dialogue box, shown in Figure 14, for defining an RGB profiles by its primaries, white point, and gamma. There is a specific entry in the list for "Monitor RGB": This does *not* reflect the profile associated with the current display, but, and I quote from the online help "This setting causes Photoshop to behave as if color management were turned off."

212

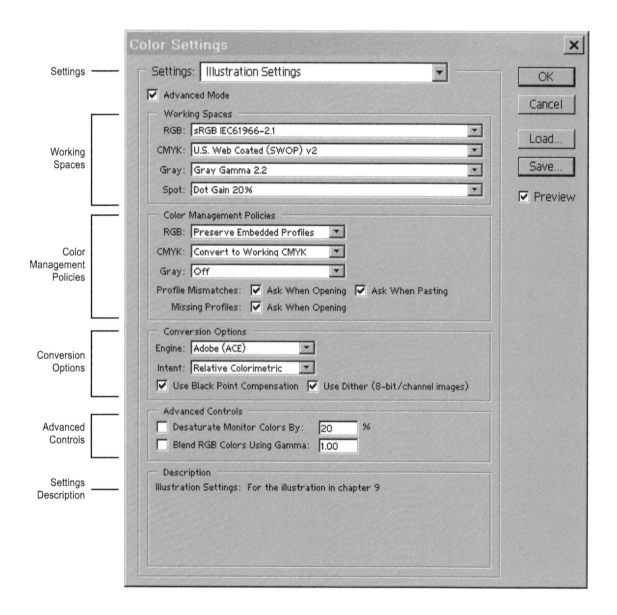

Settings

Working
Spaces

Color
Management
Policies

Conversion
Options

Advanced
Controls

Settings
Description

Figure 13.
Color settings manager from Photo-shop 6.0

Figure 14.

Control panel for creating a custom RGB space in Photoshop 6.0.

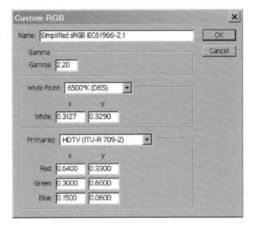

The list of selections for the CMYK space includes several international printing standards, such as "U.S. Web Coated (SWOP) v2," which translates as a profile suitable for the ink colors, gray balancing, black printing, and dot gain values common for web printing (as opposed to sheet fed) on coated (smooth, glossy) paper. The list also contains the names of many different brands of digital printers, as well as the option to load a custom profile. The "Custom CMYK" option opens a dialog box to specify the parameters for a CMYK model (Figure 15). This is recommended in the documentation for making adjustments to standard models, such as

Figure 15.

Control panel for creating a custom CMYK space in Photoshop 6.0.

changing the dot gain, but allows specifying ink colors as colorimetric values. It suggests that Photoshop does include an algorithmic model for defining print color from a combination of these parameters plus a colorimetric ink specification—the details of which I suspect are highly proprietary.

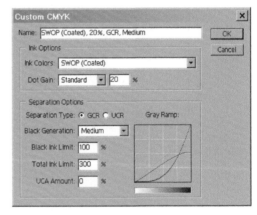

The Gray and Spot Color are one-dimensional working spaces that include parameters to describe the lightness scale. These are "dot gain" for print-oriented colors, and "gamma" for display. Spot Color is strictly a printing space, but Gray can be defined as either.

The next group of controls, labeled "Color Management Policies" defines how embedded profiles are handled. The illustration shows the three choices. "Preserve Embedded Profiles" keeps the original profile information, and does not change the pixels. This is maximally flexible, but requires the most color conversions during editing. "Convert to Working CMYK" (or RGB or Gray) performs the obvious conversion. If colors are out of gamut, they must be mapped into the current working space, which may destroy color information. "Off" means to ignore the profile information in the image, and simply copy the numeric values when editing. This is a pragmatic setting compatible with systems that don't use color management, and provides a way to ignore profile information that is wrong or inconvenient. The checkboxes in this section enable dialog boxes that can override these default policies while working.

"Conversion Options" and "Advanced Controls" only appear when the "Advanced Mode" check box (upper left corner) is set.

Under "Conversion Options" there is a way to select the CMM, called "Engine," used to perform the color transformations, and a way to select the gamut-mapping parameters. These include the rendering "Intent" and whether to match black as well as white points in the gamut mapping. "Use Dither" adds noise when converting between color spaces to reduce banding or contouring caused by insufficient number of brightness or lightness levels. This is a common problem when mapping from RGB displays to print, especially for algorithmically-generated gradations and shading.

"Advanced Controls" help with two problems when editing precisely specified RGB values; viewing working spaces larger than the display gamut (Desaturate Monitor), and accommodating both linear and non-linear working spaces (Blend RGB Color).

"Desaturate Monitor" is used when editing in a larger RGB space than can be displayed on the monitor, such as Wide RGB or Adobe RGB (Figure 10). Desaturating the primaries (moving them towards white to make a smaller triangle) can be used to make a smooth gamut mapping of the large space into the display space. The colors will not be accurate, but their relative values will be preserved, especially the brightness values that affect shading.

"Blend RGB Color" provides an explicit gamma parameter for blending. Remember that image processing, such as blending, should be performed in a linear (gamma = 1) space. However, many desktop applications blend in the non-linear display space (gamma > 1). This parameter allows Photoshop to support both practices.

The need to serve the high-end graphic arts community by including CMYK color directly adds significantly to the complexity of Photoshop's color management. Take away all of the print-specific detail, and this control panel would be much smaller, but many of the more subtle buttons, such as "Desaturate Monitor Colors" would still be needed. Supporting CMYK isn't just a concession to existing practice—it is an acknowledgement of the limitations of current device-independent color models and transformations.

# Postscript and PDF

Postscript and PDF are page description languages defined by Adobe. Support for ICC-style color profiles was introduced as part of Postscript Level 2. A Postscript printer can use the profile information included with the file to accurately render color as it prints. There can be multiple profiles referenced in a single document, corresponding to images with different embedded profiles. Like fonts, there is a tension between embedding the profile data and referencing it by name. The first is more sure, but more bulky. The color models supported by Postscript are described in detail in the *Postscript Language Reference Manual* published by Adobe Systems.

PDF is becoming a common interface between those who create formatted text and graphics and those who print it, whether on a digital printer or on a printing press. There are two basic approaches to assuring print color quality when publishing in PDF. The first is to use a well-specified RGB working space and tag all colors with this specification. Any application that creates PDF should include a way to specify the RGB working space, usually by selecting from a long list of named profiles. The sRGB color space is a common default in PC applications.

The other approach is to preprocess all colors into CMYK, using a tool such as Photoshop. In this case, color management is turned off during PDF creation, and the raw CMYK information is stored in the file. This approach requires knowing the color characteristics of the target printer, and creates a PDF file specific for that printer. It is less flexible than using a standard RGB, but allows the user to better preview the final colors. In the RGB approach, the RGB to CMYK color transformation is performed by the printer, so its results are not available until the document is printed. On the other hand, the printer may have a better conversion algorithm than that provided by a general purpose application.

# ColorSync and ICM

ColorSync and ICM are the operating system support for color management in Apple and Microsoft operating systems, respectively. Apple's solution is older and better integrated than Microsoft's, reflecting the importance of color management to graphic arts professionals, who primarily use Macintoshes.

Putting color management support into the operating systems provides a convenient way for applications to all use the same color management system and profiles, thereby giving the same results. It also provides a centralized place to put user controls, and to specify system-wide default profiles for the display, a default printer, and the default working space.

ColorSync was introduced by Apple in the early 90s. It presents an architecture for color management that allows the user to specify profiles and select which color management module (CMM) to use to implement the color transformations. It is not actually a color management system, but an infrastructure to support color management plus some OS-level controls for end-users. It provides support for embedded profiles in images, and APIs so that developers can use the system supplied color management software.

Image Color Management (ICM) was introduced with Windows 95. Like ColorSync, it provides an API and support for different CMMs, but strictly as a package for implementers—there is no system user interface. ICM 2.0 was introduced with Windows 98, and further integrated into the Windows architecture. Documentation on the Microsoft website for ICM is designed for developers. For the end-user, Microsoft has emphasized sRGB, defining it as the default color space within Windows.

In spite of ICM and ColorSync, not all applications use the color management support provided by the operating system. There is the obvious issue of legacy applications, which already perform their own color transformations. There is also a software engineering issue, where other features are considered more important than adding color management.

The ideal model of centralized color management conflicts with a practical aspect of color: the device manufacturer may do a better job of converting to and from a specific device than the centralized color management system. The profile for a printer, for example, depends on the paper and inking algorithms being used. Rather than defining a long list of different profile options, it may be simpler and more effective to let the printing system define the transformation dynamically, especially if the printer includes sensors that detect the paper and other printer settings. As a result, one valuable option for color management may be "to let the printer do it" rather than a formal profile specification for the printer. Such an option is included in ColorSync, for example.

## Summary

In an ideal color management scenario, users accurately measure and profile any digital device of interest—monitors, printers, scanners, etc., and evaluate their systems regularly to ensure quality. They rigorously define and enforce file-format transformations, both to ensure consistency and to avoid unwanted gamut mapping. In practice, this happens only in color-critical imaging industries such as special effects or graphic arts production houses. The best approach for most users is: 1) select a reasonable working space that avoids unwanted gamut mapping; 2) control those file transformations critical to their results and 3) maintain their hardware, settings, and viewing conditions so that default profiles produce acceptable results.

Digital desktop printers come with surprisingly good default profiles and algorithms for transforming from a display to a printer. All that is needed to get good results printing displayed images is an accurate display profile and the manufacturer's data for the printer. This assumes color management is turned on, and that the image colors are not too far outside the print gamut. Color management can be turned on in the print driver (PC systems), or by setting the "output profile" in ColorSync to the printer's profile.

The default profile, or profiles, for a scanner are designed for film. These profiles are not as accurate when scanning printed or painted materials, as a scanner's profile depends on the input spectra, as described in Chapter 6. Profiles are less effective for a digital camera because the input spectra are not constrained. As a result, many users treat the displayed representation, and therefore the display profile, as the precise specification for digital camera images.

Default display profiles are accurate when the display is new, but need to reflect the precise hardware settings such as those that affect the transfer curve and the white point. Anyone interested in quality digital color should create a custom profile for their display, especially if it also used as the working space for the system. It is relatively easy to improve upon the default display profile using a combination of manufacturer-supplied data and visual estimations for gamma. The Adobe Gamma Utility, which is bundled with many of their products, provides a reasonable way to establish gamma visually. Like other display profiling systems, it will adjust internal tables in the display controller to establish the desired gamma curve for all applications. Apple includes a display calibration utility as part of ColorSync. For more accuracy, packages that include a display measurement tool and software for profiling can be had for under $500.

Color management, especially gamut mapping algorithms, are still a research topic. Both the Color Imaging Conference (IS&T/SID) and the Electronic Imaging Conference (IS&T) regularly include color management, characterization, and gamut-mapping research.

Color management algorithms cannot yet match the skill of a trained professional for a specific color reproduction pipeline. They can, however, stabilize and make more predictable the star-shaped reproduction process common in digital color reproduction, saving time and minimizing frustration. For some users, this is sufficient in itself. For others, it provides a foundation for image-specific editing and adjustment that is far superior to the ad hoc results achieved without it.

# 10

# Color in Computer Graphics

In the field of computer graphics, objects and lights are virtually modeled and then rendered to create an illusion of reality, a work of art, or simply a nice picture. Once seen only in research, such imagery is now ubiquitous, especially in the entertainment industry. Color in computer graphics may precisely simulate the physics of lights and surfaces, but is more often the result of a sophisticated manipulation of RGB values. This chapter describes color in the field of 3D graphics rendering; it discusses both the synthesis of color in rendering and the reproduction of color in graphic systems.

## Introduction

Computer graphics rendering combines algorithmic creation of color, as discussed in Chapter 4, with the principles of color reproduction discussed in Chapters 5–9. Rather than capturing an image of "the real world," computer graphics synthesizes images from virtual shapes and lights, which are captured through a virtual camera. This

Figure 1.
Computer captures simulated lights and surfaces through a virtual camera to create an image.

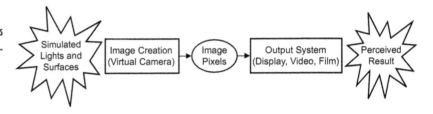

is summarized in Figure 1, which deliberately mimics the color reproduction pipeline presented in Chapter 5. In computer graphics, however, there is no visible "original" for comparison, making the correct appearance of computer-generated images even more subjective than in traditional color reproduction. Different areas in the field of computer graphics have focused on different ways to define the quality of the rendered image.

Figure 2.

A photorealistic rendering of a marble statue. (Image by Henrik Wann Jensen, Steve Marschner, Marc Levoy, and Pat Hanrahan, Stanford University. Copyright 2001 ACM, Inc. Included here by permission.)

The goal of *photorealistic rendering* is to create pictures indistinguishable from those taken of the natural world. Figure 2 is a photo-realistic rendering of a marble statue created at Stanford University. The shape of the statue was defined numerically, using data taken from a 3D scan of a real statue. The image, however, is completely synthetic. The appearance of the light, shadows, and the surface of the marble is all the result of photorealistic rendering algorithms.

In contrast, the field of *non-photorealistic rendering* was created to embrace the techniques used by painters and illustrators and to apply them to graphics rendering. For example, Figure 3 is an architectural rendering in pen-and-ink style created at the University of

*Frank Lloyd Wright's "Robie House"*

Washington. First, the authors created a 3D model to represent the building. Then, they rendered it in a style that uses pen strokes to indicate the interplay of light and shadow on the building.

Rarely are 3D graphics is neither strictly photorealistic or artistic—they are designed for some application. Computer graphics can be used to create pictures of things that are imaginary, that have yet to be built, that are too big or too small to see, that are hidden, or that are too far away. As well as being a critical component of the modern entertainment industries, computer graphics imagery is becoming a routine part of all fields where simulation and visualization of models or data are important.

Computer graphics algorithms manipulate digital representations of lights and surfaces to create an image. One approach is to simulate the physical world as accurately as possible. This requires physically correct representations and rendering algorithms, which, for color, means at minimum a spectral model of lights and surface reflectances. More often, simpler, more computationally efficient models are used, with color represented as RGB triples. These models are embedded in systems with many controls so that an experienced designer can create the desired visual effect. Modern graphics systems are a blend of physical simulation and design.

Computer graphics images must be made visible using some form of digital color output system; most commonly, this is a display. However, the image may also be targeted to print or film; creating computer-generated special effects for movies is one of the major computer graphics industries. In these cases, the pipeline in Figure 1 is really a star, as in all digital color reproduction systems. As a result, the same color management principles used in image reproduction can be applied to color in computer graphics systems.

There is a wealth of computer graphics books that describe its implementation and its application, and more are being written each day as the field expands. One of the most widely referenced books on color in computer graphics is Roy Hall's *Color and Illumination in Computer Graphics Systems*. Unfortunately, it is out of print, as well as being rather old. Andrew Glassner's *Principles of Digital Image Synthesis* is an encyclopedic reference for all aspects of rendering, and is my standard reference.

There are far more people interested in using 3D graphics than implementing graphics systems. Two of the most widely used developer's platforms are RenderMan and OpenGL. Both are available online, and have excellent references written for them: *The RenderMan Companion*, by Steve Upstill and *OpenGL Programming Guide, third edition*, by Mason Woo, et al.

This chapter starts with an overview of 3D graphics rendering systems, then proceeds to provide further detail on color specification and representation. The second part of the chapter focuses on maintaining color fidelity in graphics systems. Included in this latter section is my effort to untangle the confusion surrounding "gamma" in graphics, and the application of color management principles to graphics systems. The chapter is illustrated with computer-generated images from many sources; further information about these figures is provided at the end of the chapter.

# Rendering System Basics

The key components of a graphics rendering system are shown in Figure 4. *Object information* includes the 3D geometric models of objects in the scene, along with their color and other surface properties. *Lighting information* describes the illumination, its shape, location, and its intensity, along with its color. *Texture information* adds visual complexity by mapping images, or more generally, sample

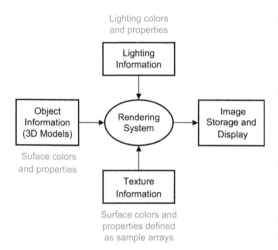

Lighting colors
and properties

Lighting
Information

Object
Information
(3D Models)

Suface colors
and properties

Rendering
System

Image
Storage and
Display

Texture
Information

Surface colors and
properties defined
as sample arrays

arrays onto the surface of objects to change the color and other surface properties. Images can also be used as environment or reflection maps, where they affect the color of light reflected from objects.

The rendering system combines these components to create an image, which is an array of RGB pixels. The image is specified by the position and orientation of a virtual camera whose optics define an image plane, as shown in Figure 5. The goal of the rendering system is to create the color values for each pixel in the image. The rendering process is optimized to create only those pixels that are visible in the image, rather than rendering the entire scene and capturing only the pixels that are visible. In effect, the scene is sampled at each pixel to determine the visible surfaces and their lighting. Image capture in computer graphics, therefore, is actually very sophisticated sampling, and the quality of this sampling is a key part of the quality of the rendering.

Figure 4.
The basic components of a graphics rendering system. Objects, lights, and textures combine to create a rendered image.

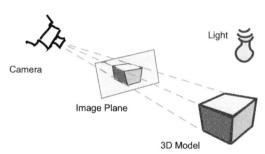

Light

Camera

Image Plane

3D Model

Figure 5.
The position and optics of the virtual camera defines the image plane.

Each point in the image—each pixel—is ultimately an RGB triple. Computing this color involves determining what surfaces are visible, what light is visible, and modeling the interaction between the two. The simulation of color in computer graphics occurs primarily in the latter two steps, called the *illumination model* and the *shading model*, respectively.

225

# Illumination and Shading Models

The basic components of lighting and shading are shown in Figure 6. Light is emitted by a light source, reflected off a surface symmetri-

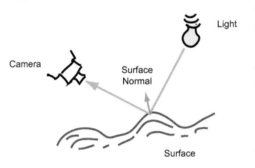

cally around the surface normal, and captured by a virtual camera. Color is generated by the shading model at the point the light interacts with the surface.

The simplest illumination models are local: light strikes the object, and is ei-

ther seen by the camera or not. Scenes rendered with strictly local illumination models don't even have shadows, though their surfaces change in brightness value depending on the amount of light received (Figure 7). Global illumination models more fully simulate the complex interplay of lights in a scene. Light reflected off one object may strike others, enabling indirect lighting as well as reflection, transparency, and shadows. Figure 8 shows two balls, one glass and one mirrored. The illumination model used in the rendering includes shadows; reflections, both in the mirrored ball and the soft, colored reflections from the walls to the floor; and refraction, which focuses light into a bright spot on the floor underneath the glass ball.

A simple shading model might produce a simple matte color plus a highlight, as shown in Figure 9. In sophisticated shading models, the surface interactions are more subtle, creating a more natural appearance. Figure 10 is a close-up of the marble statue in Figure 2, whose shading model captures the complex interplay of light both reflecting from and scattering within the stone. Shading algorithms range from quasi-physical models based on highly simplified reflection and lighting calculations, to complex surface and lighting simulations, to procedurally-defined *shaders,* such as those provided by Pixar's RenderMan.

**Figure 6.**
A light ray strikes the surface and reflects into the camera at an angle symmetric around the surface normal.

**Figure 7.**
A section of a gear system rendered with a simple, local illumination model. (Image by Eric Haines.)

**Figure 8.**
An example of a complex illumination model. (Image by Henrik Wann Jansen.)

Representing light and color in some digital form is a key component of lighting and shading models. Most of the physical models for color presented in Chapter 4 can and have been implemented to simulate the interaction of lights and surfaces in computer graphics rendering. These models require representing color physically as spectra, rays, and waves. For high-performance rendering, however, there must be an efficient transformation from the model to RGB pixels. The easiest way to do this is to simply model the colors as RGB triples, and create rendering algorithms that manipulate RGB values to give the impression of light and color in nature. Most computer graphics imagery is created using RGB as the underlying color representation. The next sections discuss color representation in more detail, both modeled and sampled.

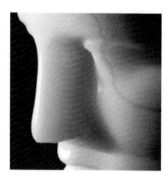

Figure 9.
A simple shading model, rendering a smooth surface with a highlight.

## Representing Color: Physical Models

A physical model of color provides the most accurate way to simulate the interaction of light, either direct or reflected, with object colors. Physical models that include wavelength information are the only way to simulate colors such as those created by dispersion, interference, and Rayleigh scattering. Physical models can also be tied to measurements of lights and materials to create a scientifically accurate simulation, as well as a quantitative method for evaluating the quality of a rendering system.

Figure 11 illustrates one of the earliest and most rigorous efforts towards quantitative evaluation of rendering systems, called the Cornell Box. In 1985, researchers at Cornell constructed a physical model of objects and lights, measured the light in an image plane with a radiometer, and compared the results to a rendered image to validate their lighting and shading algorithms. Throughout the years, several versions of the Cornell Box have been created, simulated, and measured. The box in Figure 11 is one that was constructed and studied in the late 1990s. Unlike the original Cornell Box, which contained only matte, painted surfaces, this one has shiny surfaces, including a mirror on the front face of the block in the back.

Figure 10.
Close-up of the marble surface from Figure 2, whose complex shading model includes subsurface scattering.

Reference.
Gary W. Meyer, Holly E. Rushmeier, Michael F. Cohen, Donald P. Greenberg and Kenneth E. Torrance. "An Experimental Evaluation of Computer Graphics Imagery." *ACM Transactions on Graphics* 5:1 (1986), 30–50.

Figure 11.

The Cornell Box. (a) A photograph of the physical box; (b) the rendered image. (Images courtesy of the Cornell University Program of Computer Graphics.)

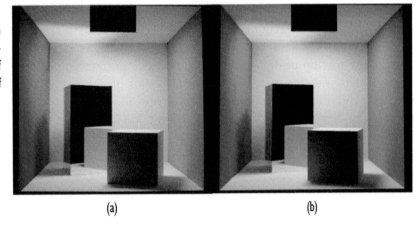

(a)　　　　　　　　　　(b)

The most basic component of a physical model for color is a spectral distribution, as described in Chapters 1 and 4. Digital models of spectra are large compared to an RGB triple. The range of wavelengths that comprise the visible spectrum runs from 370–730 nm. An equally-spaced sampling of this range, as would be created by a colorimeter or spectroradiometer, typically contains between 18 and 180 samples that must be stored for every color and light in the model and processed for every rendered pixel.

More efficient representations than uniform sampling can be created if there is some a priori knowledge of the spectral shapes. The surface reflectances of non-fluorescent objects, for example, are often very smooth, as are the spectral distributions of black-body radiators, such as an incandescent lamp. In such cases, it is possible to adapt the sampling to give a more efficient representation—the most efficient representations use as few as four samples per spectrum. A more concise representation, however, usually means a more expensive reconstruction algorithm when the spectra are used in the rendering. Various techniques have been applied to this problem, and are summarized in a paper by Roy Hall.

Another approach to efficient spectral modeling is to use basis functions, as described in Chapter 4. A set of spectra is measured and decomposed into a number of principle components. Any color in the dataset can be defined by a weighted sum of these compo-

Reference.

Roy Hall. "A Comparison between XYZ, RGB, and Sampled Spectral Color Computation." *IEEE Computer Graphics and Applications*, (1999), 36–45.

nents, so only the weights must be stored for each color. Mark Peercy applied this technique to computer graphics rendering, as is described in his SIGGRAPH '93 paper.

Reference.

Mark Peercy. "Linear Color Representations for Full Spectral Rendering." *Computer Graphics* 27:3 (1993), 191–198.

Many computer graphics systems model light as a collection of rays and trace their path throughout the scene. The simplest of these techniques, called *ray tracing*, allows the simulation of optical effects such as reflection, refraction and shadows. Add a spectral component to each ray and it is possible to model not only wavelength-specific reflection, but refraction, dispersion, and interference effects as well. Figure 12 is a computer graphics model of a prism whose base is covered with a thin film that produces interference colors. Each ray in this ray-traced rendering was split into thirty wavelength-specific components to model wavelength-specific reflection, refraction, and interference. Most of the color in the image is from the interference colors, but there are dispersion colors as well. The soap bubbles shown in Figure 13 are also examples of computer-generated interference colors.

To display spectra on a monitor, they must be converted to monitor-specific RGB values. Using trichromatic theory and the CIE color-matching functions, it is straightforward to convert from spectra to tristimulus values, as described in Chapter 1. The linear transformation that converts from CIE tristimulus values to the intensity needed for each of R, G, and B is simply the inverse of the process used to characterize an additive color system, which was described in Chapter 3.

Figure 12.

Rendering of a prism that demonstrates dispersion and colors produced by interference. (Image by Greg McNutt, Stanford University, CS 348b, 1996.)

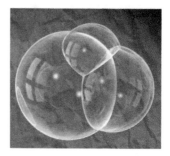

Figure 13.

Soap bubbles showing interference colors. (Image by Andrew Glassner.)

# Representing Color: RGB Values

Most color in computer graphics is represented as triples of RGB values. An RGB triple can be thought of as a tristimulus representation for a color, or as components of additive color separations. While

Reference.

Paul Strauss. "A Realistic Lighting Model for Computer Animators." *IEEE Computer Graphics and Applications* 10:11 (1990), 56–64.

individual colors can be accurately represented by such triples, their interactions at the spectral level cannot. For example, you can't generally predict the result of multiplying two spectra (such as a light with a surface reflectance) by multiplying their corresponding tristimulus values. As a result, the more spectral computations are combined, as in multiple reflections, the less physically accurate RGB is for representing the color.

RGB representations evolved by simply applying grayscale shading models derived for monochrome computer graphics, such as Phong and Gouraud shading, to each of the red, green, and blue components of a colored image. In this case, the RGB colors behave like the components of an additive display system. Such models underlie many more sophisticated shading models.

One of the most widely used RGB color/shading models was defined by Paul Strauss for the Brown Animation Generation System in the late 80s. This model evolved into the form used in OpenGL and many similar systems. It defined several color and shading parameters for the object, the light source, the highlight, and the color of objects in the shadows (ambient color), all specified as RGB triples. Its goal was to provide a more intuitive interface for the design of color in graphics than working with the raw parameters used to control graphics lighting and shading models. Figure 14 shows the

Figure 14.

The parameters for the Strauss model as used in the Virtual Reality Modeling Language (VRML), and an example of the shading created. (Redrawn view of Eric Haine's Pellucid applet.)

| Light Properties | | | |
|---|---|---|---|
| Color | R: 1 | G: 1 | B: 1 |
| Direction | X: -1 | Y: -1 | Z: -1 |
| Intensity | 0.5 | Ambient | 1 |

| Material Properties | | | |
|---|---|---|---|
| Diffuse | R: 0.2 | G: 0.5 | B: 0.4 |
| Specular | R: 1.0 | G: 1.0 | B: 1.0 |
| Emissive | R: 0.0 | G: 0.0 | B: 0.0 |

| Background Color | R: 0.3 | G: 0.2 | B: 0.3 |
|---|---|---|---|
| Gamma "Correction" | 1.1 | | |

| Shininess | 0.5 | Transparency | 0 |
|---|---|---|---|
| Shadow (Ambient) | 0.2 | | |

parameters of this model as used in the Virtual Reality Modeling Language (VRML), with an example of the resulting rendering.

The procedural shaders used in Render-Man and other systems provide the most general way to define a shading model. Rather than depend on any particular model, the designer is free to define an arbitrary procedure that blends the lighting, surface, and texture colors to create a desired effect. The default shading model is often similar to the Strauss model, but can be expanded almost arbitrarily. For example, the colors of the Morpho butterfly wings in Figure 15, whose brilliant blue-purple colors are created by interference patterns, were generated using a custom RenderMan shader. The shading algorithm modeled the wavelength-specific interaction of the light with the scales on the butterfly's wings. Note how the colors shade from blue to purple depending on the viewing angle. The foliage colors include RGB shaders plus image textures. Pixar pioneered the use of shaders for computer graphics, but many other commercial graphics systems have adopted them as well—my impression is that most commercial 3D graphics renderings are created using RGB models with procedural shaders.

Figure 16 shows the application of an RGB shading model that creates an illustrative, rather than a realistic, appearance. The model is a variation on Gouraud shading designed specifically to emphasize the shape of the mechanical model. The outlines highlighting the silhouette edges are an important part of the effect as well.

RGB representations are computationally more efficient than spectral representations. This makes it easier to create interactive systems suitable for use by a designer, who can compensate for any limitations provided by the basic model. One of my favorite stories demonstrating the inventiveness of designers involves an early graph-

Figure 15.

Butterfly with iridescent wings implemented using a custom shader. (Image by Steve Bennet and Arthur Amezcua, Stanford University, CS 348b, 2001.)

Figure 16.

A 3D model rendered to produce an illustration. (Image by Amy Gooch, Bruce Gooch, Peter Shirley and Elaine Cohen, University of Utah. Copyright 1998 ACM, Inc. Included here by permission.)

ics system that could not produce shadows, yet the designers were making pictures with shadows in them. How did they do it? They added "negative lights," which darkened the colors around them.

RGB representations are superior to spectral representations for color selection. Because vision is inherently trichromatic, manipulating RGB is more directly tied to visual response than manipulating spectra. Spectral representations are generally measured, not designed.

In summary, RGB representations are more common and flexible than physical models for color. Combined with sophisticated lighting and shading models, images based on RGB representations can look very realistic. Figure 17 is a picture based on the Cornell box that adds a glass sphere to the simple matte surfaces of the original box, and increases the complexity of the illumination model to include transparency, reflection, refraction, and caustics. The color representation is RGB, yet the picture still looks realistic. Spectral representations, however, are critical for rendering models based in simulation of physical systems, including applications like architecture and lighting design. Physical representations for color also provide a way to incorporate measured values for lights and surface properties into computer graphics rendering.

Figure 17.

A "Cornell Box" colored using an RGB (rather than a spectral) color model, but rendered with a more complex lighting model. (Image by Henrik Wann Jensen.)

## Color Images in Computer Graphics

Color in computer graphics often includes color images, or more generally, rectangular arrays of values that are used to apply texture colors, modify surface properties, simulate complex backgrounds and lighting, or even to replace the geometric model entirely. While the previous sections have emphasized modeled color, sampled color is used extensively to add visual complexity to rendered images.

Figure 18 shows a cube on a table next to a wall. A texture map has been "applied" to the cube, which means a texture image is associated with each surface, scaled and rotated appropriately. As the scene is rendered, the texture color defines the surface color at that point. The rough surface on the cube is entirely an illusion of the texture. Other texture maps have been applied to the table, the wall, and the trim. The texture maps used are shown to the right of

Figure 18.

A scene with texture maps applied to all surfaces to give them their complex colors. The textures themselves are shown to the right of the image. The same scene without textures is shown in the margin. (Image by Andrew Glassner.)

the image; the two textures on the left are photographs, the two on the right are algorithmically generated. For comparison, the small image in the margin is the same virtual scene, colored using only a simple, smooth RGB color model for the surfaces and lights.

*Bump mapping* uses an image to vary the direction of the surface normal (shown in Figure 6), which defines the direction light is reflected and scattered in the shading model. This technique gives the impression of a bumpy surface, including cast shadows, without actually having to change the model definition. The lid on the Tylenol bottle and the surfaces of the pills in Figure 19 are examples of bump mapping. The label is an image texture scanned from a real Tylenol bottle.

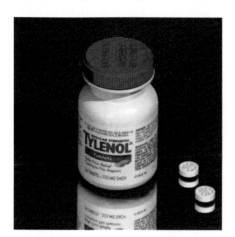

*Reflection*, or *environment mapping*, uses an image to create complex, realistic reflections. These are particularly important when rendering shiny surfaces. Figure 20 shows the application of a reflection map. The image

Figure 19.

Tylenol bottle. The label is a texture map; the embossed lid and pill surfaces are examples of bump mapping. (Image by Jeffrey DiCarlo, Stanford University, CS348b, 1996.)

(a)

(b)

Figure 20.

(a) Teapot with reflection map; (b) the image used in the map. (Images by Andrew Glassner.)

shown in Figure 20(b) can be faintly seen in the shiny teapot surface in Figure 20(a).

Complex, physically accurate lighting models are difficult and expensive to implement, especially for large environments. Using an image as a light source replaces modeling with sampling. Figure 21 shows a 3D rendering lit entirely by an image of "sampled light." That is, at each point in the rendering, the color and intensity of the light are defined by a sampled representation of intensity and color values, not a model for a light source.

The sampled light is defined by taking multiple photographs of a mirrored sphere sitting in some interesting lighting environment. The image in the lower right corner shows the sphere and the lights and colors reflected in it. These reflections are, effectively, the sampled light—they define the light shining on the sphere at that point. To give sufficient precision for rendering, several photographs taken at different exposure levels are combined to create a

Figure 21.

Hour glass lit by a high dynamic range image, captured by Paul Debevec. The sphere in the corner shows the image as it was originally captured. (Image by Brad Johanson and Jeremy Johnson, Stanford University, CS 348b, 2000.)

high dynamic range image, which has effectively floating point precision at each pixel. These images are composited together to create a sampled description of the light in the scene. Objects rendered with this sampled light look as if they are immersed in the original lighting environment.

Physically modeling all the detail in a natural scene can be an overwhelming challenge. Image-based modeling, animation, and rendering (IBMAR) replaces geometric models with volumes of sampled data. The fuzzy lion in Figure 22 is an early example of image-based modeling: it can be manipulated like a 3D model, in that it can be rotated and viewed from different angles, but has no geometric description. Instead, it is described by a correlated set of spatially arranged image samples, called a *lightfield* or *Lumigraph*. Fueled by the proliferation of affordable digital cameras, image-based techniques are blurring the boundary between a graphics model and its resulting image that was defined in Figure 1.

Figure 22.
Lumigraph rendering of a fuzzy lion. (Image by Steven Gortler, Radek Grzeszczuk, Richard Szeliski, and Michael F. Cohen, Microsoft Corporation. Copyright 1996 ACM, Inc. Included here by permission.)

## RGBA

RGB specification in graphics systems commonly includes an additional parameter, *alpha,* that indicates the degree of transparency or blending for that color. This is written as RGBα or RGBA.

A transparent color allows the color of objects behind to show through. The common interpretation of alpha maps 1 to completely opaque, and 0 to completely transparent. Figure 23 shows a blue-green shaded sphere on a red background drawn with different degrees of transparency, allowing the red to show through. The effective alpha value is included in the figure.

Different graphics systems allow different degrees of sophistication in the application of alpha, but most rendering systems support it to some degree. Most of the uses of RGB color described above, for both modeled and sampled colors, are actually implemented as RGBA.

One of the final stages of rendering is compositing rendered shapes into the final image. Alpha values are used at the edges of these fragments, to support smooth blending, a technique called

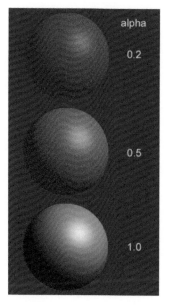

Figure 23.
Example of transparency, or alpha blending. As alpha decreases, the ball becomes more transparent.

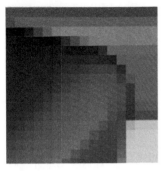

Figure 24.

Antialiasing example. Pixels at the edges of objects are blended into the background colors.

*antialiasing.* Figure 24 shows a close-up of the upper right corner of the box in Figure 18: the pixels at the edges blend into the background colors, to give a smooth appearance. Without such blending, the boundary would look "jaggy."

# Graphics Hardware

Computer graphics systems, especially for interactive, animated graphics, are often tightly integrated with the display hardware. The design of graphics hardware to support sophisticated, animated graphics has been a part of the field of computer graphics since its inception. From the flight simulators of the 80s to the latest game consoles, hardware support for the lowest levels of the rendering system has been used to make 3D animations run fast and smooth.

Hardware acceleration implements parts of the rendering system in the display controller. For example, hardware support for OpenGL might accept a stream of colored triangles from the rendering system rather than pixels, and interpolate from the vertex colors to fill the area. It is common to have hardware support for texture mapping, which again involves interpolation and placement of colors.

In recent years, graphics hardware has become extremely sophisticated, and is capable of implementing a significant part of the shading model in the hardware system. This is most visible in game consoles, where 3D graphics hardware supports rendering styles from the cheerful cartoons of Mario to the gory realism of first-person shooters like Halo, all in real time.

Color in graphics hardware is universally specified as RGBA, which is sufficient to implement transparent object colors and to allow for smooth compositing in the frame buffer.

A common part of all graphics hardware is a set of color lookup tables (LUTs) that map input pixels to output voltages. These provide a way to modify the transfer function that maps pixels to visible light without changing the display hardware. There is typically one table for each of red, green, and blue, and they generally map one 8-bit pixel value into another, although 12-bit tables are available in some systems. The practice of using such tables for "gamma correction" has caused endless debates and confusion in the graphics and display communities, as will be described in the next section.

# The Great Gamma Problem

The problem with gamma is that there are too many definitions of the term, some diametrically opposite each other. The problem became so acute that Charles Poynton wrote an article in 1998 to "rehabilitate gamma" as an effort to create some coherence out of the confusion surrounding this term. Unfortunately, the confusion persists, especially in graphics systems.

Computer graphics texts generally talk about the CRT *gamma curve*, which is the pixel-to-intensity transfer function described in Chapter 7, and about *gamma correction*. Gamma correction is used to modify the display's transfer function to give the correct appearance for rendered graphics. More generally, there are several places in the application-to-display pipeline where systems might affect this transfer function—these are essentially the same as those outlined in Chapter 9, under Specifying Color Device Transformations. The topic is repeated here in terms more specific to displays and computer graphics systems.

The transfer function that maps pixel values to intensity for a CRT is a power function whose exponent, gamma, is in the range of 2.2–2.5. Such a curve is usually referred to simply by its power, or gamma value. The solid blue curve in Figure 25, for example, is a gamma = 2.2 curve.

Rendering calculations are usually defined assuming linear intensity values; that is, numeric values in the rendering space describe a linear RGB space. To map this to a CRT display, one must

Reference.

Charles Poynton. "The Rehabilitation of Gamma." In *Human Vision and Electronic Imaging III, Proceedings of SPIE 3299,* edited by B.E. Rogowitz and T.N. Pappas, pp. 232–249, Bellingham, WA: SPIE, 1998.

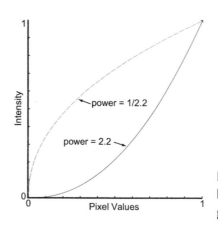

Figure 25.
Both a gamma (solid) and inverse gamma (dashed) curve.

237

apply an inverse gamma curve to compensate for the display's non-linearity. In 3D graphics, this process is called "gamma correction" because the goal is to "correct" the non-linear gamma curve of the monitor to be the linear color space defined by the rendering system. This is similar to the encoding and decoding of linear scene intensity values described in Chapter 3, in the section on Numeric Brightness Scales. In computer graphics, however, the gamma correction is usually inserted in the display pipeline using the LUTs in the display controller, rather than being applied as part of encoding the image pixels.

The dashed green curve in Figure 25 is the inverse of the blue curve, with gamma = 1/2.2 or 0.45; that is, the green curve would supply gamma correction for the blue curve. Unfortunately, it has become common to describe gamma correction curves in terms of the gamma they are correcting, or the denominator of their exponent. As a result, the description "gamma = 2.2" might describe either of the curves in Figure 25, depending on the context.

Assuming you can do a little experimenting, it is easy to tell which definition of gamma is being used. If a larger value gives you a darker picture, it is a true gamma curve; if it gives you a lighter picture, it is an inverse-gamma curve. Most gamma values presented to users are inverse-gamma.

Now, let us discuss where in the display pipeline one might encounter gamma curves of both varieties. The first is defined by the display hardware. As discussed in Chapter 7, this is normally a gamma curve like that of a CRT, even if the display is some other technology—call this the *display gamma*. The brightness and contrast controls in the display hardware can be used to change this transfer curve, however, so its precise definition depends on the hardware settings for the display. The LUTs in the display controller also provide a global place to adjust the display's appearance—call this adjustment the *LUT gamma*, though it need not be a gamma curve, as the LUTs are arbitrary pixel-to-pixel mapping tables. Call the combination of the display gamma and the LUT gamma the *display system gamma*, as summarized in Figure 26.

While PC operating systems (both Windows and Linux) normally do not modify the LUTs, both SGI (Irix) and Macintosh operating

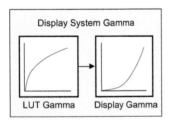

Figure 26.

The definition of the Display System Gamma, which combines the gamma created by the display hardware (Display Gamma) with that encoded in the display controller LUTs (LUT Gamma).

|  | PC | Mac | SGI |
|---|---|---|---|
| Default Display Gamma | 2.2 | 2.2 | 2.2 |
| Default LUT Gamma | 1.0 | 1.22 | 1.7 |
| Display System Gamma | 2.2 | 1.8 | 1.29 |

Table 1.
Default display system gamma for
three graphics platforms.

systems do modify the LUTs to create a characteristic default display
system gamma. This is why pictures designed on one platform are
consistently too dark or too light when viewed on another. Table 1
summarizes the default gamma values for each platform. To match
convention, the LUT gamma is described by its inverse, so the display
system gamma equals the display gamma divided by the LUT gamma.

While the default LUT Gamma is set by the operating system,
there are usually controls that allow users to modify the LUT
values. Most often, these controls are in terms of an inverse gamma
curve. For example, on an SGI system, typing "gamma 2.2" will
put a gamma = 1/2.2 curve in the LUTs, creating a linear pixel-
to-intensity mapping (display system gamma = 1.0). Similarly,
many PC display controllers present a graphical UI that allow
the user to modify brightness, contrast, and gamma by changing
the LUTs. These controls are usually found under the Advanced
Features of the system Display Properties. The Monitor Control
Panel in the Macintosh, however, asks the user to specify the total display system gamma rather than directly manipulating the
gamma correction.

Display characterization programs also modify the LUTs to create a desired display system gamma, usually 2.2 for the PC and 1.8
for Macintosh systems. In these applications, the user is asked to
define the target display system gamma, and the program modifies
the LUTs to create the overall desired result. To determine the display gamma, the user either measures the display transfer function,
or uses visually evaluated test patterns to estimate it.

Arbitrary applications can also change the LUTs, though this is
very bad form from a systems perspective, as it will change the ap-

Figure 27.
Two paths for creating the desired viewing space for rendered graphics. The top line changes the Display System Gamma, the bottom line does not.

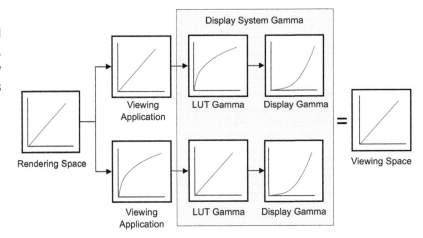

pearance of all applications and invalidate any characterization created for the display. References to "LUT Wars" in characterization software suggests, however, that this practice is not as rare as it should be.

Image and graphics display programs, such as "xv" for X-Windows, often offer an option to apply a gamma correction value for viewing the image. This is applied by the application as it writes the image colors to the display controller, and does not affect either the LUTs or the image pixels. Therefore, it is a strictly local effect, which is seen only when the image is viewed in that specific application.

Figure 27 shows two different pipelines of "gamma" settings. The top line uses the LUTs for gamma correction, the bottom line applies it in the viewing application. Both create the same visual results, but the top line changes the display system gamma for all applications.

Image editing applications allow the user to apply gamma curves to images, permanently changing the pixels. Again, these are usually specified as inverse gamma curves—a larger gamma value creates a lighter, less saturated image.

Gamma correction for computer graphics is classically applied in the LUTs, as illustrated in the top path in Figure 27. The original graphics systems were stand stand-alone applications, and as a result, older graphics textbooks recommend using the LUTs for gamma correction. In multi-application systems with multiple windows, however, the gamma correction is usually applied by the application displaying the graphics, as shown along the bottom path in Figure 27. The problem with this approach is that the image will look much different displayed through the graphics application than in other imaging tools, which more commonly use the default, non-linear display system settings. A more robust approach would be to apply the gamma correction as the image is generated, encoding the pixels non-linearly, as is common for cameras and other image capture systems.

# Managing Color in Computer Graphics

The intense focus on displays and display hardware in computer graphics can obscure the applicability of device-independent color specification and management to this field. Trying to print or project a carefully rendered image, however, makes the need for color management painfully clear.

The solution is the same as for color image reproduction; establish an RGB working space, and control the transformations in and out of it. The degree of control and precision in the specification and the transformations will define the degree of cross-media color accuracy. Companies creating computer graphics special effects for movies and television all use some form of color management, which is usually a proprietary implementation, not ICC profiles and commercially available CMMs, but the principles are the same.

Figure 28, which is repeated from Chapter 9, shows color management centered around a device-independent working space. For computer graphics, this is most commonly the same space

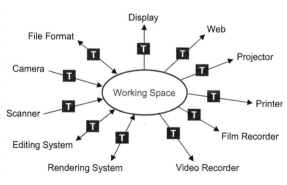

Figure 28.

Device-independent color management for computer graphics is centered on a standard working space. Each purple transformation box must be carefully specified and controlled.

241

as that used by the rendering system, which is a linear RGB space. As in image reproduction, there are two basic approaches for defining the colorimetric properties of this working space.

The first is to use a display-centered working space. In this case, the primary colors and white point of the RGB working space match those of the display system, but describe a linear RGB space to optimize for rendering and compositing. This makes the transformation between the working space and the display hardware simply the change in intensity transfer function, or the "gamma correction" described earlier. Classically, the gamma correction was implemented in the display hardware, so the working space and the display space were identical.

The second approach uses a device-independent RGB working space that is large and precise enough to include all the colors of interest. In this case, the display is just another output device, which may not accurately show all the colors in an image.

The relative strengths and weaknesses of these two approaches are the same as described for color image reproduction in Chapter 9. A display-oriented system is more convenient because all colors are visible and the display is an intrinsic part of the user's work flow, but the color specification is restricted to the display's gamut, and may not include all the colors available on some other medium such as film. The second approach solves the problem of the limited color space, but for any combination of image and output, there may be colors out of gamut.

A company that creates solely computer graphics imagery, such as Pixar, can (and does) use the display-centered approach. This is also the most convenient approach for most computer graphics researchers and for those strictly interested in output to DVD or video. A company that routinely combines CG with live action film, such as Rhythm & Hues, must use a wider RGB space to encompass the film gamut.

Colors provided as input to the rendering system must also be precisely defined with respect to the working space to give a consistent result. Colors are most often specified as RGB triples for both models and textures. Colors may also be specified as spectra, in

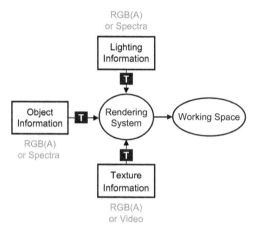

RGB(A)
or Spectra

Lighting
Information

**T**

Object
Information

**T**

Rendering
System

Working Space

RGB(A)
or Spectra

**T**

Texture
Information

RGB(A)
or Video

which case they must be converted to RGB. Figure 29, which combines concepts from Figure 28 and Figure 4, shows the different color specification paths and transformations.

Figure 29.

Colors input into the rendering system must be transformed to the working space.

By default, all RGB values should represent points in the RGB working space. Similarly, image colors used for texturing should also represent linearly-encoded RGB values in the working space. Many images, however, are captured by cameras that non-linearly encode the color values to make more efficient use of the 8 bit encoding, as was described in Chapter 6. As a result, such images must be converted to linear RGB, ideally without introducing contouring and other quantization artifacts. A full RGB-to-RGB color transformation may also be necessary if the primaries for the standard space are significantly different from those of the image-encoding primaries.

Image textures are often created interactively using an image editing system such as Photoshop. What do the RGB pixel values created by such a system mean? The pixels are displayed during editing; therefore, the corresponding physical values are the ones viewed on the display, which are defined by the display characterization as was described in Chapter 3. Ideally, these should be accurately mapped into the working space before being used in the rendering system. Pragmatically, it is often sufficient to convert them from non-linear to linear values, by mapping them through the display system gamma. This applies also to colors selected for lights and surfaces using a color selection tool.

Spectral representations for color are first converted to tristimulus values, then converted to the standard RGB working space using the processes described in Chapters 1 and 3. All values are intrinsically linear with respect to intensity, so this process is simply a mat-

243

ter of integration and matrices. However, the resulting colors may be out-of-gamut, or may not give the correct appearance for reasons akin to exposure control and white balancing when capturing natural scenes with a camera.

For example, equal energy illumination (a flat spectral distribution) is a common, convenient default light source for graphics rendering. Unless "white" in the working space is exactly the same color (approximately 6,000 K), images rendered using this illumination will have a slightly bluish or slightly yellowish tint. If the goal is precise, colorimetric simulation, than this result is correct. If the goal is to create a good-looking image, then mapping white to white, as is common in photography and image reproduction, will give a better result. This mapping can be achieved in the XYZ-to-RGB matrix, or applied as a separate, more perceptually defined process. Alternatively, this problem can be addressed when targeting the image defined in the working space to a particular display or other medium. In this scenario, the transformation is more like gamut mapping.

## Example: Color Management at Rhythm & Hues

The following is a summary of the Rhythm & Hues color management system, which was presented at SIGGRAPH 2002 as part of the Courses program. It illustrates all of the principles above, except for the problems introduced by using spectrally-defined colors.

The standard RGB color space at Rhythm & Hues is a high-precision, linear space. When stored in files, the colors are encoded logarithmically for compactness. The standard color space can be mapped to the appropriate color values for display, and to the film and video production processes. These processes contain several steps, but the characterization is defined for the entire process, from rendered pixels to measured final media. Intermediate steps are stabilized via process control. For example, film creation is characterized from the values written to the film recorder to measurements on the final print. Intermediate steps that affect the color, such as film developing, are not modeled, but are standardized using sensitometry and other process control techniques.

Reference.
Pauline Ts'o. "A Field Guide to Digital Color: Color at Rhythm & Hues." Course 21, SIGGRAPH 2002, July 2002.

In some cases, computer generated imagery must match live-action footage, which is scanned and composited with the computer-generated effect. In this case, the pixels from the film scanner must be mapped to the standard space. This is done by a combination of metric and visual comparisons, using test samples taken directly from the film being scanned.

Images imported for textures must also be characterized to map accurately into the standard space. Often they are generated by artists using digital painting tools on workstations. These values are transformed through the display system gamma to compensate for the difference between the non-linear workstation RGB values and the linear working space values. No effort, however, is made to match the specific display primaries.

As in all computer graphics, the display is the primary output medium while the graphics are being designed. The final results, however, must ultimately be tuned for their appearance on film or video, depending on the project. Different mappings are used to simulate film or video on the designer's display. Because of the difference in color and dynamic range between a display and projected movie film, it is not usually possible to completely replace previewing on film by viewing on a display.

# Calculating Luminance

The color chapter of computer graphics texts often includes a discussion on converting RGB to luminance, which is correlated with the perceived brightness of a color (see Chapter 2). This is useful for computing luminance contrast, which is key to sharpness and legibility, and for converting full color images to grayscale.

In many graphics texts and tutorials, an equation derived from the NTSC definition of luma (the Y of IYQ), is routinely recommended to compute luminance for any arbitrary RGB triples. This is wrong. Not only is this based on an obsolete set of phosphors, it confuses the non-linear luma with luminance (see Chapter 6). It should be clear to readers of this book that the correct equation depends on the RGB characterization of the targeted display, and cannot be universally defined.

To compute luminance from an RGB pixel, the pixel values must be mapped to linear intensity values (r, g, b). The luminance of the pixel is defined as:

$$Y = rY_R + gY_G + bY_B,$$

where the constants are the luminance values for each of the R, G, and B primary colors defined by the RGB specification. Different RGB primaries will create different luminance equations. Changing the white point will change the relative magnitudes of the primaries, which will also change the luminance equation.

Assuming the pixels represent colors that are going to be displayed, the constants in the luminance equation are defined by the display primaries, and the display system transfer function is used to map the pixels to linear intensity.

If, for some reason, it is impossible to obtain the actual display system characteristics, the ITU Rec-709 primaries used in digital video and in sRGB offer the most reasonable approximation for modern monitors (though not for flat panel displays). It is even more important to correctly define the intensity transfer function of the display system. The default values in Table 1 can be used as a starting point, although it is easy for any individual display system to deviate significantly from those values.

The equation for luminance depends only on the color and brightness of the primaries. For a display, these are defined by the phosphors and the white point. Using as an example an sRGB monitor, with its standard phosphors and D65 white point (D65), the values for $Y_R$, $Y_G$, and $Y_B$ are the same as those shown for Y in Table 2 in Chapter 3 (page 60). The resulting equation for luminance is:

$$Y = 0.2126r + 0.7152g + 0.0722b.$$

Using the same phosphors but a white point of D93, which is the default for many desktop CRTs, the luminance equation becomes:

$$Y = 0.1799r + 0.7216g + 0.0985b.$$

These values have been normalized to give a maximum luminance of 1.0. To generate a physical value for luminance, they must be scaled by the luminance of white, which is 80 candellas/meter$^2$ for sRGB.

# Tone Reproduction in Computer Graphics

As in all image reproduction systems, controlling the tone reproduction is an important component of computer graphics rendering. Models for tone reproduction in natural image reproduction compare the original scene luminance values to the reproduced values. In computer graphics, the definition of the "original" is less clear. If the rendering system is based on a physical model, color and brightness have physical meaning, which can be mapped to the physical values produced by the output display. More commonly, all lights and colors are simply RGB(A) values, which have no physical meaning independent of the display system.

In image reproduction, a camera captures a natural scene, converting it to pixels (perhaps via film and a scanner), which are then further processed to create a visible result on a display or print. In computer graphics, rendering algorithms replace the camera, creating an image of (theoretically) unbounded RGB values. These values need to be compressed into the physical display gamut, usually represented as a unit RGB color cube. The process is often called *tone mapping.*

The classic method of tone mapping in graphics is to scale the image values by a constant amount to fit them into the desired range. There are various ways to chose the scale factor, from simple ones based on the min/max or average value of gray, to complex, perceptually-based algorithms. Any pixels too colorful after tone mapping are simply clipped. If the scaling and clipping do not produce an acceptable image, the best solution is often to change the original scene description, adjusting lights and object colors to avoid the problem, rather than applying more complex tone reproduction models.

References.

Jack Tumblin and Holly Rushmeier. "Tone Reproduction for Realistic Images." *IEEE Computer Graphics and Applications* 14:6 (1993), 42–48.

Sumananta N. Pattanaik, James A. Ferwerda, Mark D. Fairchild, and Donald P. Greenberg. "A Multiscale Model of Adaptation and Spatial Vision for Realistic Image Display." In *Proceedings of SIGGRAPH 98, Computer Graphics Proceedings, Annual Conference Series*, edited by B.E. Rogowitz, pp. 287–298, Reading, MA: Addison-Wesley, 1998.

Work by Jack Tumblin and Holly Rushmeier published in 1993 introduced the use of visual principles to control tone reproduction for computer graphics renderings. Their method incorporates models of human perception, and creates renderings that reflect the absolute brightness of the image values; that is, a room lit by a dim light still looks dim compared to one lit by a bright light. Simple scaling would normalize both images to the same appearance. Such algorithms require physical units to define the absolute brightness of the colors, and are most obviously applicable to physically-based rendering systems. Their work has stimulated further research in incorporating perceptual models as part of the rendering pipeline. One framework for this approach was published by Pattanaik, Ferwerda, Fairchild and Greenberg.

There has been a flurry of work in the last few years in the computer graphics community on tone reproduction algorithms for scenes with extremely high dynamic range. Globally scaling such scenes leaves them looking either too dim or too bright, and with regions lacking detail. Better algorithms use non-linear mappings, some of which vary adaptively to different portions of the image. Figure 30 shows the difference between a linearly scaled image and one created by applying an adaptive algorithm based on the principles used by photographers. The associated paper (see the Notes about the Figures section) contains a good summary and examples of different tone reproduction algorithms developed in computer graphics over the last decade.

Figure 30.

A high-dynamic range image linearly scaled in RGB (a) versus adaptively scaled (b). (Images by Erik Reinhard, Michael Stark, Peter Shirley and Jim Ferwerda. Copyright 2002 ACM, Inc. Included here by permission.)

(a) (b)

Research performed at Xerox PARC in the early 90s, called Device-Directed Rendering, tried to blend gamut-mapping techniques (as used in color management systems) with information from the rendering system about how the pixels colors were generated. The goal of this work was to automatically adjust light and object colors to keep all rendered colors in-gamut, rather than applying tone-mapping or  gamut-mapping algorithms to the image pixels. The specific approach proved to be too inefficient to be practical, but the concept remains sound— to take advantage of the algorithmic nature of computer generated imagery to perform intelligent adaptation to the target medium.

Reference.
Andrew Glassner, David Marimont, Ken Fishkin, and Maureen Stone. "Device-Directed Rendering." *ACM Transactions on Graphics* (1995), 58–76.

# Summary

Computer graphics blends color synthesis with image reproduction. A computer graphics scene description consists of virtual objects and light sources, which must be rendered to create a visible image. Their colors may be derived from physical models of lights and surfaces, but are more often a matter of clever manipulation of simple RGB(A) color values with complex lighting and shading algorithms. There exists a wide range of rendering styles, from photorealistic to purely artistic. Photorealistic images created with computer graphics now rival those captured from nature.

Twenty years ago, creating images by rendering 3D models was practiced in only a few universities, or as a part of custom-built engineering design systems. Even the simplest images took hours of computing time, and were primarily of technical, rather than visual, interest. Now, such simple pictures can be animated on a desktop computer. Computer-generated special effects, from alien monsters to talking pigs, from surrealistic landscapes to the Titanic, are intrinsic components of movies, television, art, and advertising. Computer graphics is also used in science, engineering, and design. It is becoming a routine part of all fields where simulation and visualization of the results are important.

Color in computer graphics has several roles. It is input into the rendering system to describe the color of lights and objects. Here, color can be physically modeled as spectra, rays, and waves, or simplified to RGB(A) values. Color also appears in images that are used to define surface textures or environment maps. Sampled colors are almost always RGB(A) values.

The output of the rendering system is also a colored image, which must be reproduced in some physical form to become visible. In graphics, the output medium is most often an additive display. Older graphics texts bind assumptions about displays and display systems tightly into their model of graphics and rendering algorithms. In the broader context of digital color reproduction, however, the output of a computer graphics rendering system should be specified with respect to a device-independent RGB color space, which is then mapped to some specific color output device.

Using a device-independent color model for computer graphics provides a foundation for on-line publishing, cross-media reproduction, and for accurate display. A primary goal of this chapter is to make the computer graphics community aware of the principles of color management and how to apply them to rendering systems. These principles and their application have already been discovered by those using computer graphics to create cross-media special effects.

The same principles can be used to ensure accurate, repeatable results for all graphics systems. All too often, rendered images that are the result of thousands of precise calculations on colors and lighting include textures of completely unknown color and linearity, and are evaluated on monitors of unknown characteristics. If there is to be any science in the evaluation of computer graphics rendering, there need to be standards that control how color is specified and viewed. Thanks to desktop color management systems, the tools for color measurement and display characterization are now available and affordable. Those involved in designing and creating computer graphics imagery, whether commercially or in research environments, should use them.

# Notes about the Figures

The examples of computer graphics rendering in this chapter come from various sources. While these are acknowledged briefly in their captions, this section provides more information and references.

Some examples of simple shading, textures, and environment maps (Figure 18, and Figure 20) were created specifically for this chapter by Andrew Glassner. Andrew also created the soap bubbles in Figure 13, which are excerpted from an image he had previously rendered for his column in *Computer Graphics and Applications* (November, 2002).

I created the various simple shaded spheres using Eric Haine's Pellucid applet, which is also the basis for the illustration in Figure 14. The applet can be found at

http://www.acm.org/tog/resources/applets/vrml/pellucid.html

Eric Haines also gave me the example of a simple shading model in Figure 7, which I excerpted from an image he has used in his own talks on global versus local illumination models.

Henrik Wann Jensen donated two examples of global illumination models, for Figure 8 and Figure 17, as well as contributing to the rendering of the marble statue in Figure 1.

Several images (Figures 12, 15, 19, and 21) were created by students at Stanford University as part of the annual rendering competition. This competition is organized as part of CS348b, which is titled "Computer Graphics: Image Synthesis Techniques." The students are challenged to model and render a real object, and the results are judged by a panel of experts on both their technical accuracy and their aesthetics. The winners of each competition are awarded an all-expenses paid trip to the annual SIGGRAPH conference. The student figures are labeled with the students' names and the year they entered the competition. For further information on the rendering competition, visit:

http://graphics.stanford.edu/courses/cs348b-competition/

The Cornell Program for Computer Graphics provided the Cornell Box images. Further information about the box in its various forms is available on the Cornell website:

http://www.graphics.cornell.edu/online/box

The following figures were published previously. Here are the full citations to those publications:

Figure 2. Henrik Wann Jensen, Steve Marschner, Marc Levoy, and Pat Hanrahan. "A Practical Model for Subsurface Light Transport." In *Proceedings of SIGGRAPH 2001, Computer Graphics Proceedings, Annual Conference Series*, edited by E. Fiume, pp. 511–518, Reading, MA: Addison-Wesley, 2001.

Figure 3. George Winkenbach and David H. Salesin. "Computer-Generated Pen-and-Ink Illustration." In *Proceedings of SIGGRAPH 94, Computer Graphics Proceedings, Annual Conference Series*, edited by Andrew Glassner, pp. 91–100, New York: ACM Press, 1994.

Figure 11. Sumanta N. Pattanaik, James A. Ferwerda, Kenneth E. Torrance, and Donald Greenberg. "Validation of Global Illumination Simulations through CCD Camera Measurements." In *Proceedings of the IS&T 5th Color Imaging Conference*, pp. 250–253.

Figure 16. Amy Gooch, Bruce Gooch, Peter Shirley, and Elaine Cohen. "A Non-Photorealistic Lighting Model for Automatic Technical Illustration." In *Proceedings of SIGGRAPH 1998, Computer Graphics Proceedings, Annual Conference Series*, edited by Michael Cohen, pp. 447–458, Reading, MA: Addison-Wesley, 1998.

Figure 22. Steven J. Gortler, Radek Grzeszczuk, Richard Szeliski, and Michael F. Cohen. "The Lumigraph." In *Proceedings of SIGGRAPH 96, Computer Graphics Proceedings, Annual Conference Series*, edited by Holly Rushmeier, pp. 43–54, Reading, MA: Addison-Wesley, 1996.

Figure 30. Erik Reinhard, Michael Stark, Peter Shirley, and James Fewerda. "Photographic Tone Reproduction for Digital Images." *ACM Transactions on Graphics* 21:3, 267–274.

# 11
# Color Selection and Design

Selecting a color and applying it to a document, diagram, or web page is an act of color design familiar to most computer users. The process is simple: pick a color and apply it. Creating an aesthetic and functional result, however, is difficult. In the visual arts, color design focuses on the relationship among colors, as this dominates the appearance of any individual color. Digital color selection tools, however, focus on picking colors one at a time—rarely do they support any coordinated selection of groups of colors. Good color design is traditionally taught by example and experience. It is guided by the principles of color harmony, and constrained by functional and practical limits dictated by perception, convention, materials, and cost. This chapter discusses color selection first as it is taught in design, then as it commonly appears in digital systems, along with some of the functional and perceptual principles that influence color design.

255

# Introduction

Color design is traditionally the domain of artists and designers. In any designed object, be it a painting, advertisement, or automobile, the colors were selected by some human being for their visual effect. The color design presented in this chapter is primarily 2D color design, as presented in graphic arts, although the basic principles can be transferred to other domains, including 3D graphics and visualization.

The principles that underlie color design are called *color harmony*. This chapter summarizes the basic principles of color harmony, but the principles alone are insufficient—experience and examples are crucial for creating a good color designer. Wucius Wong's *Principles of Color Design* presents a systematic approach to color design that many will find very clear and compelling, with many lovely examples. Josf Albers' *Interaction of Color* is a classic book on color design and perception organized as a set of exercises designed to explore different color relationships.

The principles for digital color selection were established back in the late 70s, and, unfortunately, have changed little since then. Two different reorganizations of the RGB color cube, HSV (same as HSB) and HLS (same as HSL) were defined to give a more "intuitive" way of selecting colors, and they form the basis for most color selection tools. These tools offer little support for color design principles, but are highly interactive, which allows skilled users to compensate for their inherent limitations.

For those already familiar with color design, but less familiar with digital color, Anne Spalter's book, *The Computer in the Visual Arts,* has a good chapter on digital color for designers, as well as a wealth of other information. While targeted at the visual artist, it is useful reading for anyone wanting to bridge the gap between visual design and digital systems.

This chapter first introduces the basic principles of color design, including some examples. It then describes the world of digital color selection, including the underlying schema for HSV and HLS, the RGB-based color spaces used in most color selection tools.

256

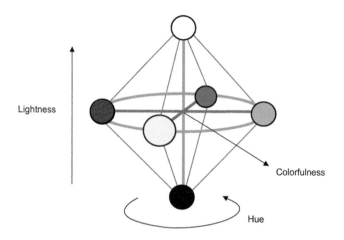

Figure 1.
The perceptual organization of color, which matches a designer's organization as well.

It also discusses selection methods that are more perceptually based, including the use of color names.

# Principles of Color Harmony

Designers define color by its hue, saturation, and value (lightness/darkness), matching the terminology of perceptual color spaces. Color in design is organized along these three dimensions: *value* runs from dark to light on the vertical axis, *saturation* increases radially, and *hue* is defined as an angle. This matches the perceptual organization of color introduced in Chapter 2, which is shown again in Figure 1, relabeled to match the terminology above. In fact, Wong uses precisely the Munsell terminology described in Chapter 2 (hue, chroma, and value) to describe color design.

Good design focuses attention using *contrast*, and unifies the design using *analogy*. Analogous features are similar, where as contrasting ones are different, as illustrated in Figure 2. Contrast and analogy in the color domain can be applied to hue, saturation, or value.

Most discussions of color in design start with a hue circle, as shown in Figure 3. The hue circle used by designers is formed from three primary colors—red, yellow, and blue—plus three secondary colors created by "mixing" the primaries pairwise—orange, green,

Figure 2.
Differences in hue, value, and chroma create contrast; similarities create analogy.

257

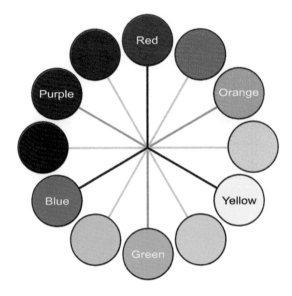

and purple. The full circle contains all the adjacent, pairwise mixtures of these six colors as well, as shown in the figure. Analogous colors lie close together on the color wheel, and contrasting ones lie opposite. The *complement* of a color lies opposite it in the color wheel, and defines the highest contrast in hue.

A similar hue organization is used for paint mixture. A color mixed with its complement should become neutral, or gray. Of course, the actual neutral achieved will depend on the specific paint colors and their composition.

Those familiar with printing often wonder how an artist's red, green, and blue primaries relate to the cyan, magenta, and yellow used in subtractive media. Visually, cyan could be called blue and magenta red, and some old-time printers do so. Figure 4 shows the printing primaries used in this book organized as a designer's hue wheel. The colors are not the same as shown as in Figure 3, but demonstrate that the mixture principles are similar. Note that printing colors are "mixed" by layering transparent inks (which act like dye layers), rather than blending them as in paints. The colors in Figure 3, which are reproduced as CMYK colors, were visually selected and blended.

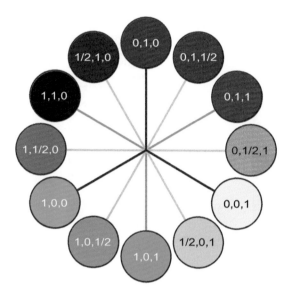

Figure 4.
Hue circle created from printers inks, with primaries magenta, cyan, and yellow. The ink coverage is shown for each patch, where 1 = solid ink.

Value describes the perceived lightness of a color. The simplest value scale is a grayscale, or set of achromatic colors organized by lightness. Any hue can vary in value, and colors of different hue can also be ordered by value. Figure 5 shows an achromatic value scale, and one that changes color. Value is proportional to the color's luminance, and the Munsell dimension "value" is mathematically correlated with $L^*$, as described in Chapter 2.

Contrast in value is key to the definition of shapes and edges—it is the critical factor for text legibility and for the perception of sharpness. A good design should remain legible when all the colors are reduced to shades of gray proportional to their value.

A chroma scale starts out vividly colored, then fades to a gray of the same value, as shown in Figure 6. It is very difficult to vary chroma without varying value, and vice versa, unless one has a very carefully calibrated color output system or a very trained eye. While Figure 6 was designed to vary only in chroma, by the time it is printed, it will most likely vary slightly in hue and value also.

More commonly, designers think in terms of gradations and mixtures that don't lie precisely along the perceptual dimensions. The term *tint* applies to a color that has been lightened and desaturated

Figure 5.
Two color blends, from dark to light. Both illustrate the same value scale.

Figure 6.
A blend from a high-chroma color to gray. Ideally, only chroma changes, not value or hue.

259

Figure 7.

Tints, tones, and shades of five saturated hues.

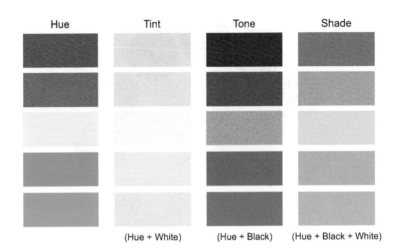

| Hue | Tint | Tone | Shade |
|---|---|---|---|
| | (Hue + White) | (Hue + Black) | (Hue + Black + White) |

by mixing it with white. A *shade* is a color grayed and darkened by mixing it with black. A *tone* is a color mixed with both black and white. Examples are shown in Figure 7.

Figure 8.

Contrast decreases as the background becomes lighter. Without sufficient contrast, the text is illegible.

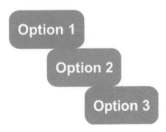

Figure 9.

To give the options the same emphasis, the three colored backgrounds have the same value.

# Controlling Color Value

Controlling value is key to good design. Legibility and sharpness are defined by contrast in value. Different hues with equal value are a common design element for selection buttons, tabs, and other indications of equivalent categories.

Perceptually, edges are defined only by value contrast. This can be described quantitatively as luminance contrast. ISO 9241, part 3, recommends a minimum contrast ratio of 3:1, with 10:1 preferred, and a ratio of 5:1 is often quoted as a safe rule-of-thumb. Figure 8 demonstrates how contrast is important for legibility. Even though the hue contrast remains high, once the value contrast drops, the text is illegible.

Creating different hues of equal value is also important in a design. Figure 9, for example, shows three colored blocks with white text, representing equally important options. To make the text equally readable, and for the options to all appear to have the same importance, they should have the same value, as shown.

Luminance can be computed for a color display from the luminance values of the primary colors, as described in Chapters 3 and 10. The accuracy of the luminance calculation depends on the accuracy of the transfer function, phosphors, and white point used to construct the equation. However, few color design tools that offer a translation from color to "luminance" (gray) are using a display-specific transformation. Only those that support color management and have an accurate profile for the display have the data to do so.

The perceptual value L* used in the CIELAB color space was designed to match the Munsell value scale. Therefore, systems like Adobe Photoshop that support the use of the CIELAB color space (often called simply "Lab") provide a direct way to determine and define value, assuming that color is being managed accurately.

Both luminance and L* are quite easy to compute from the information in a display profile. Now that profiles and color management systems are becoming more common, it should be easier to include tools in digital design systems to define, evaluate, and constrain value.

## Color on Color

Placing one strong color next to another induces a number of perceptual effects that can strengthen or destroy a design. The first is simultaneous contrast, as described in Chapter 2, and illustrated again in Figure 10. The blue-green rectangles are all exactly the same color, but appear different because of the surrounding colors.

Another psychophysical phenomena that affects the appearance of color is caused by chromatic aberration in the optics of the eye. The effect of chromatic aberration is that focus varies with wavelength, so different colors focus at a different depth in the eye. As a result, some colors recede and others appear to float with respect to their background colors. This is shown in Figure 11, where various colored squares are displayed on both an intense red and blue background. Different colors float and recede with respect to the two different backgrounds. This effect is much stronger on a monitor, where the colors are closer to being pure spectral colors. Try, for example, putting pure red text on a pure blue background. The red

Figure 10.
Simultaneous contrast makes the identical cyan squares appear quite different.

261

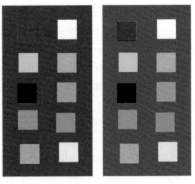

Figure 11.

Different focal points for different colors make them appear to float or recede relative to their background color. The color patches are the same on both backgrounds, except that red and blue are swapped in the upper left corner.

stands out dramatically in front of the blue.

The eye must make subtle adjustments to accommodate a difference in focus. If there is an edge between two colors that requires a significant change of focus, the edge may actually appear to vibrate, especially if the luminance contrast is also low. Figure 12 may vibrate in this manner, depending on the printing process. Continuously changing focus is also tiring.

It is difficult to focus on edges created with very saturated blue colors that stimulate primarily the short wavelength cones. These cones are more sparsely distributed in the fovea than the medium and long wavelength cones (by more than a factor of 20). This sparse distribution, combined with the difference in focus, can make very pure blue text and fine lines difficult to read. This effect is difficult to demonstrate in print, but is far too easy to do on a monitor: simply color the text RGB = (0, 0, 1) and display it on a black background. The result is very unpleasant. Because of this problem, one should avoid intensely blue colors for text and other fine detail, especially on a black background. A much better bright blue is (0, 0.5, 1), which is both easier on the eyes and easier to reproduce.

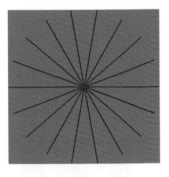

Figure 12.

It is difficult to focus on both the red lines and the blue background due to chromatic aberration.

In summary, predicting the effect of color on color is difficult. The simplest solution is to avoid it entirely, layering colors only with neutrals, as in colored text on a gray or white background, or white text on a darkly colored background. One can then move cautiously from neutrals to muted colors. At all times, it is important to control the value difference between the foreground and background colors, to ensure that edges are visible and text is legible.

# Color Blends

A common process in design is to blend two colors to make a smooth gradation. The simplest and most common blends combine a color

with black or white, as shown in the top two stars in Figure 13. To look smooth, a blend must seem to change continuously, without any visible jumps from one digital color step to the next. A visible jump will introduce contouring or banding, as shown in Figure 14. Blends are often the stress case for a digital color output system, exposing problems with the linearity or number of quantization levels that are invisible in more complex images.

Blending colors of two different hues is more complicated—how do the colors mix? Do they mix like paint, like printing ink, or like colored light? To the implementer, a blend is an interpolation path in some color space. Obviously, different spaces will give different results, as the paths contain different colors. For displays, a simple interpolation in RGB gives a reasonable blend, as shown in the bottom two stars in Figure 13; but if the target is print, it is safer to blend in CMYK to avoid flattening and contouring caused by mapping out-of-gamut RGB colors. Figure 15 was carefully constructed to show the type of problems caused by gamut mapping. On a display, it is a smooth ramp from bright blue to white.

Here are some technical notes on how Figure 15 was constructed: The original gradient was computed in RGB, running between pure monitor blue (which is severely out-of-gamut in print) and white. It was then converted to CMYK, which included gamut mapping to the print gamut. The dark blue part of the gradient shows a long sequence of no change due to projection to the gamut surface. There is also some hue shifting in the midtones. Back in Chapter 2, when introducing CIELAB, there was reference to problems with hue uniformity, especially for saturated display colors. The hue shifting in this image illustrates those problems.

## Color Palettes

If you go to the graphic arts section of your local bookstore and look for a book on color design, most of what you will find are books of color schemes; pages upon pages of pre-designed color combinations, typically two harmonious and one contrasting color. These colors, plus their tints and tones, are used to define the color palette for a

Figure 13.
Stars colored with blends or gradients. Clockwise from the upper left: red to black; blue to white; cyan to green; and orange to yellow.

Figure 14.
The orange to yellow star, severely quantized to illustrate banding or contour lines. Normally, quantization effects are more subtle.

Figure 15.
A gradient constructed in RGB may not be smooth when gamut-mapped to CMYK—this is an extreme example.

263

Figure 16.

Different color palettes applied to the same design.

design. Having a large collection of different color schemes gathered together in a book makes it easy for designers to visually consider and evaluate many different combinations.

Different color schemes convey different messages. For example, a palette of warm red and yellow tones is more vibrant than a cool blue one. A bright, saturated color scheme suggests youth; a subdued one sophistication and maturity. Figure 16 shows the application of different palettes to a common design. Even in this simple example, the different palettes have significantly different visual impacts.

Many harmonious color schemes have a predictable geometric relationship with respect to the hue wheel. The simplest is a monochromatic scheme, which is simply a single color and its tints and tones. The middle color scheme in Figure 16 is a monochromatic scheme. An analogous color scheme consists of several similar colors, as shown in Figure 17(a). A contrasting, or complementary color scheme takes colors opposite each other, as shown in Figure 17(b). A split complementary scheme combines a color with the two colors bracketing its complement, as shown in Figure 17(c). Note that in most cases, the hues are not included in their pure, saturated form; e.g., a red/green color scheme (Figure 17(b)) doesn't have to look like a Christmas card.

There have been research tools for selecting color palettes based on these geometric interpretations of color harmony. Giordanno Beretta's MetaPalette applied the same geometric arrangements described above to CIELAB space to select palettes of harmonious colors. This work influenced the Canon Color Advisor, a clever tool buried in the print driver of Canon color printers.

References.

G. Beretta. "Color Palette Selection Tools." *SPSE's 43rd Annual Conference, The Society for Imaging Science and Technology* (1990), 94–96.

L. Lavendel and T. Kohler. "The Story of a Color Advisor." *Sixth IS&T/SID Color Imaging Conference* (1998), 228–232.

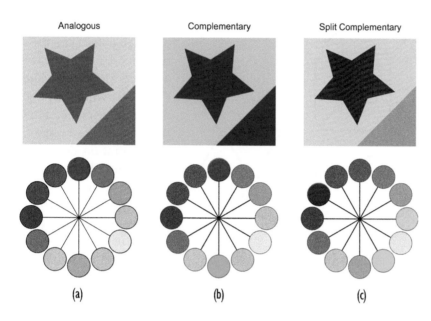

Analogous       Complementary       Split Complementary

(a)          (b)          (c)

Figure 17.
Different color schemes, together with their geometric relationship to the color wheel. (a) Analogous; (b) complementary; (c) split complementary.

Given the importance of palettes in color design, it is surprising how poorly they are supported in digital color design and selection. Many systems allow the user to store selected colors, but few make it easy to design, modify, and reuse palettes. Even professional design tools, like Adobe Illustrator, treat palettes more as collections of color selections, rather than as the basis for a color design.

# HSV and HLS

HSV ("hue, saturation, value") and HLS ("hue, lightness, saturation") are remappings of the RGB color cube that provide a more perceptual or artistic organization of the color cube. First defined in the late 70s, they have become ubiquitous components of RGB color selection. Both HLS and HSV begin with the following observation: Take the RGB color cube and stand it on its diagonal, with black at the bottom and white at the top. Its projection is a hexagon with the corner colors at the vertices, as illustrated in Figure 18. HLS and HSV are piecewise

References.
Alvy Ray Smith. "Color Gamut Transform Pairs." *Computer Graphics* 12:3 (1978), 12–25.
"Status Report of the Graphics Standards Planning Committee." *Computer Graphics* 13:3 (1979).

Figure 18.

The projection of the RGB color cube is a hexagonal "hue circle" when black and white are aligned.

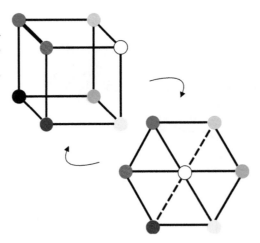

linear reorganizations of the RGB color cube such that one dimension (L or V) traces the black-white diagonal, where all the neutral colors lie. Hue steps around the hexagon, and saturation describes the distance the color lies from the neutral axis.

HSV (also called HSB) was originally designed to mimic the painter's tints, tones, and shades. Starting with a fully saturated color, decreasing S literally adds white light to the color. Decreasing V darkens the color, or "adds black." In this way, tints, tones, and shades can easily be constructed in HSV. While these are not true paint color mixtures, this model provides a simple way to think about the dimensions of this space. The HSV color space describes a hexagonal cone with black at the tip and white in the center of the flat face. White and all of the primary and secondary colors have the same value, V = 1, even though their perceived lightness is quite different.

Table 1.

Different measures of lightness for the primary and secondary colors of an sRGB display. Y and L* are perceptual measures of lightness, V and L from HSV and HSL are not.

|  | Y | L* | V (HSV) | L (HLS) |
|---|---|---|---|---|
| red | 21 | 53 | 1.0 | 0.5 |
| green | 72 | 88 | 1.0 | 0.5 |
| blue | 7 | 32 | 1.0 | 0.5 |
| cyan | 79 | 91 | 1.0 | 0.5 |
| magenta | 28 | 60 | 1.0 | 0.5 |
| yellow | 93 | 97 | 1.0 | 0.5 |
| white | 100 | 100 | 1.0 | 1.0 |

HLS was designed to mimic a perceptually organized color system, and is structured as a double-ended cone. It has the same hexagonal cross-section as HSV, but the primary and secondary colors are organized at the middle of the cone, where L = 0.5. HLS is closer in shape to a perceptual space, reflecting that colors near black and white generally have a more limited range of saturation available, and that colors near white have higher L values than the primaries and secondaries. However, L is still only a measure of the strength of the RGB signal—it does not reflect the color's true luminance value. Table 1 shows values for luminance (Y, normalized to 100), L* (from CIELAB), V (from HSV), and L (from HSL) for each of the primary and secondary colors, assuming an sRGB display.

The precise algorithms for HLS and HSV are presented in Anne Spalter's book, as well as in most computer graphics texts and tutorials. One or both are included in most graphics packages.

# RGB Color Selection

One aspect of digital color familiar to almost all users of computer systems is the color selection tool. A typical desktop system might have over a dozen different tools for selecting RGB colors embedded in its applications. This section talks about the design of these tools, illustrated with specific examples. In all cases, my analysis is based on observation, rather than specific knowledge of the tool's implementation.

There are a few ideas that consistently appear in RGB color selection tools. The 3D space is mapped to a 2D interactive tool, usually some combination of 1D and 2D sliders. The RGB color cube is often first transformed into either HLS or HSV, as these are considered more "intuitive" than manipulating RGB values directly. There is a color patch to show the current color, which may be combined with a color patch showing the starting color. Most tools offer numeric feedback for the color, which can be edited or typed in directly.

Figure 19.
R, G, and B sliders. This tool can also be configured to show CMYK, HSB, and Lab sliders. (Adobe Photoshop.)

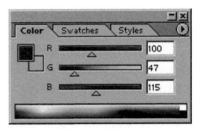

The simplest interactive tool presents sliders for each dimension such as R, G, and B or H, S, and V, as in Figure 19, which shows a set of RGB sliders from Adobe Photoshop. This tool allows a choice of sliders (RGB, CMYK, HSB, and Lab). It also has a miniature palette of colors along the bottom for quick selection, a common addition to many color selection tools. The palette is organized as a "spectrum," with light colors at the top and dark ones at the bottom. There is a choice of palettes—this is the one convenient for selecting CMYK colors.

Many color selection tools map two dimensions to a plane, then includes a linear slider for the third. Figure 20 shows the elegant layout for HSV used by MetaDesign (previously Fractal Design) Painter. There is a circular slider for controlling hue surrounding a triangular slider that controls both S and V together. Moving towards the black corner of this triangle "adds black," moving towards the white corner "adds white." The fully saturated color is in the third corner. The complete color selection tool includes HSV sliders and type-in boxes as well. The white and purple boxes in the lower left corner are the foreground (purple) and background (white) colors.

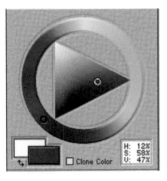

Figure 20.
The layout for HSV used by MetaDesign Painter. The hue circle surrounds an S–V triangle. S = 0 is the white corner, V = 0 is the black corner.

The Adobe Photoshop Color Picker is shown in Figure 21. This tool augments the slider tool shown in Figure 19. It can be configured in a variety of ways. Figure 21(a) shows Hue assigned to the linear slider, with S and V (or B) arranged in a rectangle; move left to decrease S, down to decrease V. The numeric values for all four of the color spaces supported by Photoshop are visible in the tool, and can be edited directly. The small number at the bottom is the hexadecimal code for the color as used in HTML. It indicates R, G and B as two hex digits (eight bits) each. Figure 21(b) shows the tool configured for CIELAB, with L* on the slider, a* and b* on the rectangular slider (a* is horizontal).

268

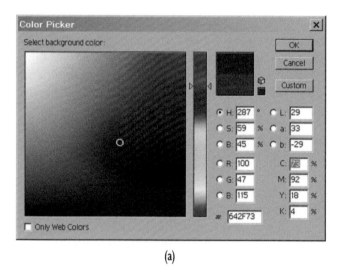

(a)

(b)

The tool shown in Figure 22 is a color selection tool provided as part of the Microsoft Windows development environment. This screen shot is from Microsoft Office; it shows the HLS space, with L (mislabeled luminance) on the slider and an H–S slider. Figure 23 shows a different arrangement of HLS from Jasc PaintShop Pro, which is similar to the arrangement provided by Painter (Figure 20). The hue circle surrounds an L–S slider, with L on the vertical axis. In this arrangement, it is easy to see that L runs from white to black, with the saturated colors halfway between.

Figure 21.
The Adobe Photoshop Color Picker, which can show many different configurations and color spaces. (a) A linear hue slider and a rectangular S–V (or S–B) slider; (b) a linear L* slider and a*–b* slider showing the same color as (a).

Figure 22.
The Microsoft Windows HLS color picker with a linear L slider and an H–S slider.

Figure 23.
The arrangement of HLS provided by Jasc PaintShop Pro, with an S–L slider surrounded by a full hue circle.

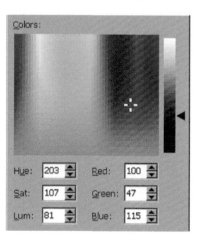

269

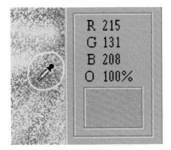

Figure 24.

The eyedropper tool (circled), which picks up the color beneath it. (Jasc Paintshop Pro.)

Possibly the most important digital color tool is the eye-dropper, shown in Figure 24, which allows a user to copy a color from elsewhere in the design. This, plus the highly-interactive nature of digital color tools, makes them unique for color design. Being able to quickly and easily experiment helps compensate for the inherent limitations in the tool design and the underlying color spaces.

# Perceptually Based Color Selection Tools

One of the original applications of perceptually based color systems like the Munsell book of color was color selection. There have been some efforts to use perceptual spaces as the basis for digital color tools as well. One of the first to be commercially available was TekHVC by Tektronics. The TekHVC space is similar to CIELUV, and the TekColor Picker could show slices of the gamuts of two different devices, making it possible to select only common, in-gamut colors, as shown in Figure 25. This system vanished as Tektronics left the color business, and has been supplanted by the use of CIELAB in device-independent color systems.

The advantage of a tool based on a perceptual space is that it is easier to select colors with similar perceptual properties, such as dif-

Figure 25.

A screen shot of the TekHVC color selection tool, showing a slice of both a print and a monitor gamut. (Image courtesy of Tektronics.)

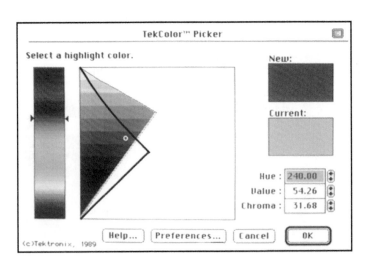

270

ferent hues of the same perceived lightness or saturation. The three dimensions are perceptually independent, unlike those of the HLS or HSV systems.

A disadvantage of using a perceptual, rather than a device-specific color space, is that a specific device gamut is an irregular volume inside the color selection space. For example, Figure 26 shows an sRGB display gamut in CIELAB space, which is quite a different shape than the simple, hexagonal cones of HLS and HSV. This makes it more difficult to create a simple interface that constrains the user to select only displayable colors. Constrained navigation may leave the user "stuck" in sharp corners, such as the red corner on the left side of Figure 26. The alternative is to simply flag out-of-gamut colors, rather than constraining the user's actions. However, this makes it harder to select colors at the gamut surface, such as solid (no halftones) colors for printing.

Reference.
J.M. Taylor, G.M. Murch, and P.A. MacManus. "Tektronix HVC: A Uniform Perceptual Color System for Display Users." *SID 88 Digest* (1988), 77–80.

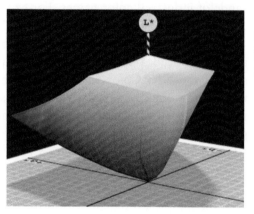

Figure 26.
Monitor gamut in CIELAB, showing its irregular shape. (Image by Bruce Lindbloom.)

# Color Names

There is something fundamental about naming color. Ask someone to describe a color and they will begin first with a hue name such as blue or green, then modify it with adjectives such as dark, light, vivid, bright, etc. Berlin and Kay, in their book *Basic Color Terms*, reported that color names are common to human language, and that there is a standard set of 11 simple color names that appear in all mature languages. These are the perceptual primaries (red, green, blue, yellow, black, white, and gray) plus orange, purple, pink, and brown (Figure 27). Furthermore, the order the names are introduced into the language is consistent across languages, and further study has found that there is a

Figure 27.
The 11 basic color names identified by Berlin and Kay.

high degree of correlation between the names given to colors across people and cultures. That is, the color patches described as "red" by one person will also be described as red by another. Of course, not all colors have simple names, and there is less agreement on how these are specified, but the agreement for the basic colors is very good.

Color names are also important for industrial applications. In 1976, the National Bureau of Standards published *A Universal Color Language and Dictionary of Names*, by Kelly and Judd, which describes both an orderly language for specifying color, calibrated to the Munsell color space, and a mapping into this language of thousands of industrial color names. For example, the color "avocado" could be described as "dark olive green."

Berk, Brownston, and Kaufman published an article in 1982 that introduced a simplified, more computationally tractable version of the NBS Color Language. We had an implementation of their system (plus our own improvements) running at Xerox PARC in the 80s. A commercial implementation of the NBS Color Language is presented as part of the Color Science Library from the CGSD Corporation in Palo Alto, CA (www.cgsd.com).

In the graphic arts domain, the Pantone Matching System is a common way of describing colors. Originally published as a book of print samples, the digital form of this system is incorporated in most of the major graphic arts applications such as Adobe Illustrator or Quark Express. The *Pantone Guide to Communicating with Color* is a book of color designs specified as Pantone colors.

Dictionaries of color names that bind a text string to an RGB triple are used for specifying colors in HTML, in X-Windows, and surely in other applications as well. I have some reservations about the names themselves (what color is "silver," anyway, on a computer monitor?), and they are rarely selected in any organized manner that would aid in aesthetic design. However, the idea of using names for such specifications is a good one. Names can be used to limit the color palette, ideally to one that has been carefully designed. They also abstract the color from its precise specification in a natural way. For example, one could specify three distinct hues, (e.g., blue, green, and purple) and have the program that displays

Reference.
T. Berk, L. Brownston, and A. Kaufman. "A New Color-Naming System for Graphics Languages." *IEEE Computer Graphics and Applications* 2:3 (1982), 37–44.

them select specific RGB triples based on the display characteristics. This could be as simple as selecting colors available in a limited color map, or as complex as choosing colors based on perceptual or aesthetic constraints.

An additional layer of abstraction names colors by their function, such as background, title, and highlight. Then, color schemes can be assigned to these functional descriptions. Microsoft PowerPoint, for example, provides this functionality for designing slides.

# Summary

Color design is a delicate art that involves balancing colors for both aesthetic and functional reasons. The concept of color harmony is defined by Wong as "successful color combinations, whether these please the eye by using analogous colors, or excite the eye with contrasts." In art and design school, the basic principles of hue, saturation, value, contrast, and analogy are taught by example and by experience. Excellent color design requires skill and experience, but there are concrete principles that can be applied by anyone, or even applied algorithmically. Value can be defined by luminance or $L^*$; legibility can be defined in terms of luminance contrast. The subtleties of using color on color can simply be avoided. Predefined color palettes are useful to everyone, especially when modulated by blends, tints and tones of the key colors. Color schemes can also be defined geometrically, ideally in a perceptual color space; but even in the simpler spaces defined by HSV and HLS, analogous colors lie closer together and contrasting ones lie farther apart.

In the digital color world, color selection is almost exclusively about picking points out of a color space. The ability to pick colors by navigating a seemingly infinite space of possibilities has appeal, especially to the technical mind, but aesthetic color design is all about constraint, which is why palettes are so popular for professional designers. Digital tools provide little support for managing palettes and other collections of related colors, and it is not easy to evaluate or constrain colors for their function, such as creating a set of hues of equal value.

Reference.

Barbara J. Meier, Anne Morgan Spalter, David B. Karelitz. "Interactive Color Palette Tools." To be published in *Computer Graphics and Applications*.

Color names are another way to abstract and constrain color selection. Color names are a basic property of perception and language, and are used in industry in systems such as the Pantone color naming system. Using a name can provide a level of abstraction in the color specification that can be used to make the color more robust.

With a little planning and a bit of trickery, interactive color selection allows a designer to rapidly create many different color schemes for a design. For example, Photoshop is often used to mock-up designs, such that all elements of the same color appear on the same layer. This makes it fairly easy to select everything on a layer and recolor it, to test different color schemes. There is no automatic way, however, to move from this representation to the final one; color values must be copied and applied by hand. As another example, some designers evaluate the luminance contrast of different colors by switching their display into grayscale mode. Unfortunatly, their tools do not make it convenient to do this in software. All of this rather strongly suggests that some new digital color selection tools, based on designer's needs, could and should be created.

Recent work at Brown University by Barbara Meier, Anne Spalter, and David Karelitz has resulted in the design of a suite of color selection tools based on design principles, including user-defined palettes based on geometric relationships in the Munsell color space and several innovative tools for creating and manipulating them. The suite allows the user to rough out the color composition before focusing on the form, and provides analysis tools for color frequency. The tools have been implemented as plug-ins for Adobe Illustrator, and hopefully point the way towards more design-oriented color selection tools.

# 12
# Color in Information Display

Look around and consider what role color plays in the objects you see. I would argue that one of its primary purposes is to inform. Even in the natural world, color is a key way to tell if something is edible, dangerous, or one of a related set of objects. In the manufactured world, color is used extensively to label, to discriminate, to brand, or to otherwise identify objects and their components. It is an important part of illustration, cartography, and data presentation, from medical imaging to pie charts. This chapter describes the informative aspects of color in fields called visualization, information visualization, and illustration. Its principles also apply to the use of color in user-interface design.

## Introduction

Color is a key component in information display. When used correctly, it can make a complex task or analysis simple; when abused, it can turn a simple presentation into visual chaos.

Visualization as a computational field explores ways to present visually the result of measurement and simulation. From simple charts and graphs, to complex, multi-dimensional models; from fields as diverse as cartography, engineering, business and medical imaging—color is one of the parameters manipulated as part of information and data display.

Color in user-interface design follows the same principles as color in visualization—the color is used to identify components of the interface. Contrasting colors call attention, analogous (similar) colors group and blend pieces together. Making components legible and unambiguous is a critical part of user-interface design, and the right color design is a key part of this process.

Edward Tufte has published three marvelous books on information visualization, but only one, *Envisioning Information,* has a chapter on color. Tufte describes the fundamental uses of color as follows: *to label* (color as noun), *to measure* (color as quantity), *to represent or imitate reality* (color as representation), and *to enliven or decorate* (color as beauty). These principles provide the framework for this chapter.

Many of the positive (and negative) effects of color in visualization have a perceptual basis. Colin Ware's book, *Information Visualization; Perception for Design,* describes many of these effects for color as well as for other perceptual phenomena.

Most great visualizations, such as those shown in Tufte's books, are designed by skilled, even inspired, experts. The challenge for the field of computer-generated visualization is to assist ordinary people in creating such visualizations, or ultimately, to create such imagery automatically. It is unrealistic to expect the results to be as wonderful as the best of those designed by professionals, but they should be effective at conveying the desired information, and, at minimum, visually inoffensive.

## Good and Bad Uses of Color

The goal in selecting colors for information display is, ultimately, to produce an image that is attractive and that conveys its message effectively. Anyone planning to use color in information display should thoroughly understand the principles of color design pre-

sented in the previous chapter. Color should be clarifying, not con-
fusing. It should be tasteful, and it should be robust across media,
viewing conditions, and viewers.

Tufte's primary rule for the use of color is "do no harm," as color
used poorly can be worse than no color at all. Color can cause the
wrong information to be most prominent, can cause text and other
features to be illegible, and when overused, give the effect of many,
unrelated pieces "screaming" for attention (sometimes called the
Times Square, Los Vegas, or Piccadilly Circus effect, depending on
where you live). Figure 1 is a simple example designed to show these
problems. Coloring each letter a different color (Figure 1(a)), makes
the words nearly unreadable. Some letters pop out, others recede,
and the "e" seems more associated with the frame than with the rest
of the text. Identically colored letters may cluster, even in ways that
cross word boundaries. Those that are too light are illegible. Coloring each word separately, as shown in Figure 1(b), respects the word
structure, but now "careful" is emphasized over "color" (which may
be appropriate). Figure 1(c) puts both words in red, where they stand
out from the blue frame. Figure 1(d) uses the same blue as the frame,
giving the most unified design; now, the color is entirely aesthetic—
it could be black and read the same.

Map makers are masters of the effective use of color. Figure 2 is an
example from the U.S. National Park Service showing the west end of
St. Thomas in the Virgin Islands. This single example demonstrates all
of Tufte's principles. Color is used extensively to label, as indicated by
the legend (color as noun). The shape and height of the terrain is suggested by the shading (color as quantity); water is blue, modulated by
greenish sea grass; the land is brown, accented by a green mangrove
stand (color as representation). The map is also pleasing to look at,

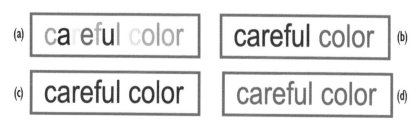

Figure 1.
Color creates emphasis and group-
ing. (a) Incoherent application of
color to text obscures the words;
(b) "careful" is emphasized; (c) the
message is separate from the frame;
(d) unified text and frame.

Figure 2.

Map of St. Thomas, demonstrating multiple uses of color. (Courtesy of the National Park Service.)

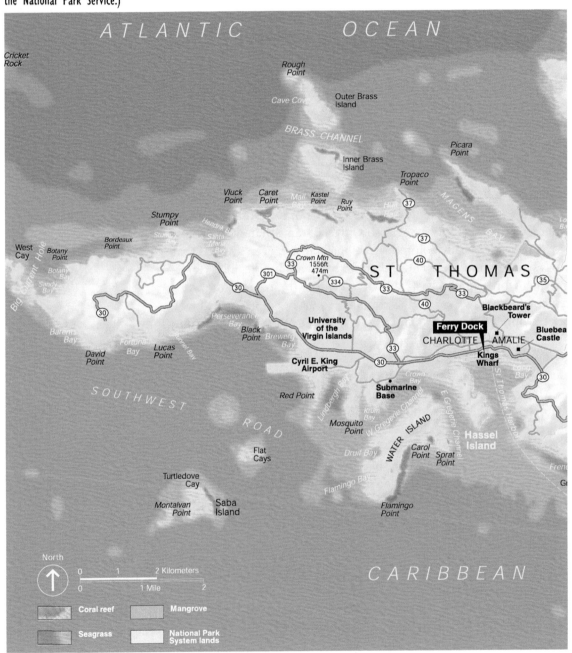

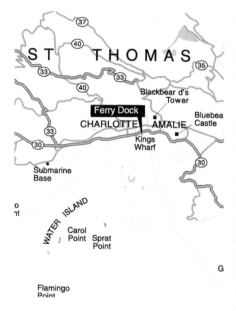

with colors chosen to harmonize as well as inform (color as beauty). Figure 3 shows a fragment of this map rendered in shades of gray. While the roads and labels are legible, it is much less effective, and much less beautiful, than the full-color version.

Computer-generated visualizations are created from data. Figure 4 shows an MRI section of a head from the National Institute of Health's Visible Human Project. Here, data taken in slices are visualized to reconstruct the original shape. In this grayscale representation, the result is recognizably a cross-section of a human head, with identifiable structures such as the brain and the sinus cavities.

There is no color information captured in an MRI; it simply captures the different densities of the materials scanned. The grayscale visualization maps the density scale to a lightness scale to create the image such that denser material is lighter (to mimic the appearance of x-ray film). Coloring the image in Figure 4 would mean replacing the grayscale values with colors, a process called *pseudocolor*. Such coloring is of dubious value, unless there is a clear mapping between a specific density and a feature of interest. Two pseudocolorings are shown in Figure 5. The blue-yellow gradient is visually inoffensive, possibly even appealing, but it is doubtful whether it conveys more information than the grayscale image. It has a similar lightness scale, even though it also changes hue. The brightly colored "rainbow" mapping is, alas, the more usual application of pseudocolor. Not only is it garish, notice how several small yellow and green "features" have artificially appeared around the cheeks. It is hard to imagine any form of visualization that would be improved by such a coloring—unfortunately, it is the default for many visualization systems.

Figure 3.

Fragment of Figure 2, reproduced in shades of gray only. While legible, it is much less effective than the colored version.

281

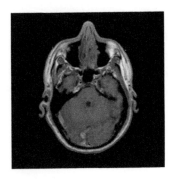

**Figure 4.**

MRI scan through a human head. The shades of gray indicate density, with white being the most dense.

This chapter focuses on effective uses of color as label and quantity. Labeling is the most fundamental use of color, as it can discriminate and cluster related features. Color as quantity is more complex. Much that is written about color and visualization concerns effectively mapping quantitative values to color. In many applications, this means finding color sequences, or *color scales*, to represent numeric values. Color as quantity also includes the use of color and shading to represent shape and size, as in the MRI example above, and also in 3D models and renderings, as will be shown below. Such visualizations are an application of the computer graphics principles (described in Chapter 10) to visualization.

Finally, color should be robust. Information presented in color should be visible to those with color vision deficiencies and to eyes of all ages. If rendered on multiple media, it must be legible on all. This implies both careful design and good color management.

**Figure 5.**

Two pseudo-colorings of the image in Figure 4. (a) A blue-yellow gradient, which creates a similar value scale and (b) a "rainbow" map, which obscures the original data.

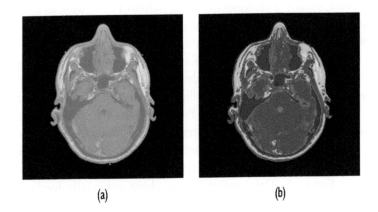

(a)  (b)

## To Label

Color can be used to identify or label objects in an illustration. Figure 6 shows a map of the area around Point Reyes, California. Color is used extensively to identify each feature. The major roads are red; the lesser roads are black. The area within the national park is green;

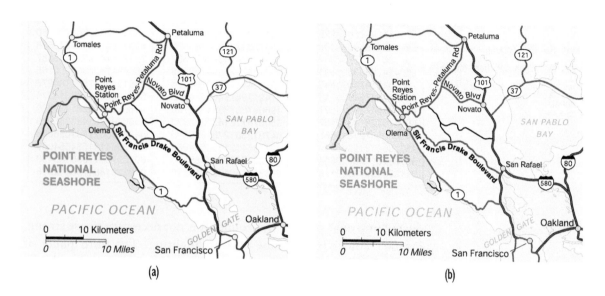

(a)                                                                (b)

Figure 6.

Color as label. The major roads (in red) are much easier to identify in the colored map than the gray version.

the water is blue—in both cases, the text is similarly color-coded. Cities are yellow, outlined in black to provide a contrasting edge to the light color. The major highways (101, 80, 580) are also outlined in black, which makes their red color appear darker, an example of simultaneous contrast. Figure 6(b) shows the same map in shades of gray. While legible, it is clearly much less effective without the labeling effect of the colors.

Color's ability to label is a low-level perceptual phenomena— you don't have to think about it. It is a classic example of a preattentive process; the visual system separates the colored objects from their background all at once, early in the visual processing. This can be demonstrated with following classic experiment. Take a page of numbers, such as shown in Figure 7, and count all the 7s. The amount of time to count them is proportional to the number of digits on the page. That is, to count the 7s, you must look at each digit to determine whether or not it is a 7. Now, color the 7s some distinctive color such as red, as in the second panel in Figure 7. In this case, the time it takes to count the 7s is proportional to the number of 7s on the page, independent of the number of other digits. This phenomenon is sometimes called *pop-out*, as the colored

Figure 7.

Count the 7s. Coloring them red makes them "pop out" from the surrounding numbers, and much easier to count.

| | | |
|---|---|---|
| 7348572647568799860 | 7348572647568799860 | 7348572647568799860 |
| 6947264785934848696 | 6947264785934848696 | 6947264785934848696 |
| 7847367410293635587 | 7847367410293635587 | 7847367410293635587 |
| 9504947825364809165 | 9504947825364809165 | 9504947825364809165 |
| 7381343547502184676 | 7381343547502184676 | 7381343547502184676 |
| 5749129475462514375 | 5749129475462514375 | 5749129475462514375 |
| 4960976572351432750 | 4960976572351432750 | 4960976572351432750 |
| 6506787261433245279 | 6506787261433245279 | 6506787261433245279 |
| 6476125612781056895 | 6476125612781056895 | 6476125612781056895 |
| 4672354121534654987 | 4672354121534654987 | 4672354121534654987 |
| 6072376142385385490 | 6072376142385385490 | 6072376142385385490 |
| 0163201864798012878 | 0163201864798012878 | 0163201864798012878 |

digits "pop out" of the background. It is not necessary to use strong, primary colors for labeling and highlighting. The third panel of Figure 7, in which the 7s are much less vividly red, still exhibits pop-out.

Color also works well to group related objects, either by coloring them similarly, or by placing them on a colored background. Figure 8 shows an array of names and numbers representing a number of measured tristimulus values for different projected colors, which are measured for several different projectors. Coloring alternate rows makes it easier to associate each color name with its data, especially as the rows are long. Coloring sets of columns, as shown in the lower table, clusters all the measurements for each projector.

| | X | Y | Z | X | Y | Z | X | Y | Z | X | Y | Z |
|---|---|---|---|---|---|---|---|---|---|---|---|---|
| red | 25.37 | 13.70 | 0.05 | 26.27 | 14.13 | 0.04 | 18.41 | 10.16 | 0.05 | 17.43 | 9.30 | 0.00 |
| green | 22.14 | 51.24 | 0.35 | 20.68 | 49.17 | 0.44 | 21.11 | 46.00 | 0.20 | 16.36 | 37.95 | 0.12 |
| blue | 13.17 | 3.71 | 74.89 | 15.38 | 5.20 | 86.83 | 11.55 | 3.37 | 65.53 | 9.96 | 3.44 | 56.14 |
| gray | 63.46 | 73.30 | 78.05 | 64.66 | 71.99 | 90.08 | 52.96 | 62.49 | 67.99 | 45.54 | 53.65 | 58.14 |
| black | 0.66 | 0.70 | 0.77 | 0.63 | 0.66 | 1.09 | 0.47 | 0.58 | 0.70 | 0.44 | 0.54 | 0.71 |

| | X | Y | Z | X | Y | Z | X | Y | Z | X | Y | Z |
|---|---|---|---|---|---|---|---|---|---|---|---|---|
| red | 25.37 | 13.70 | 0.05 | 26.27 | 14.13 | 0.04 | 18.41 | 10.16 | 0.05 | 17.43 | 9.30 | 0.00 |
| green | 22.14 | 51.24 | 0.35 | 20.68 | 49.17 | 0.44 | 21.11 | 46.00 | 0.20 | 16.36 | 37.95 | 0.12 |
| blue | 13.17 | 3.71 | 74.89 | 15.38 | 5.20 | 86.83 | 11.55 | 3.37 | 65.53 | 9.96 | 3.44 | 56.14 |
| gray | 63.46 | 73.30 | 78.05 | 64.66 | 71.99 | 90.08 | 52.96 | 62.49 | 67.99 | 45.54 | 53.65 | 58.14 |
| black | 0.66 | 0.70 | 0.77 | 0.63 | 0.66 | 1.09 | 0.47 | 0.58 | 0.70 | 0.44 | 0.54 | 0.71 |

Figure 8.

Color clusters either the rows (top table) or sets of columns (bottom table).

Color highlighting and labeling is most effective if there is a small number of different colors on a relatively neutral background. Otherwise, the display becomes cluttered and the information it presents becomes difficult to find. The principles of good design suggest that 2 to 4 key hues, plus tints, tones, and shades of the basic palette will provide a visually coherent design.

If color labeling is used such that the color uniquely conveys the information, it is subject to all the limitations of human memory. The rule of thumb for short term memory is 7, plus or minus 2. Note, however, that an experienced user could memorize many more mappings as long as they are distinguishable when seen in isolation. Colors used in this way should have distinct names. It is probably safe to assume that most people remember the color name more readily than the actual hue, and when discussing visualization, it is helpful to be able to name the visual attributes of the features of interest, such as "the red ones." Color labeling that conflicts with information presented more explicitly, such as text, should be avoided. Figure 9 illustrates the "Stroop Effect," which illustrates a fundamental conflict between reading and color labeling. Try saying out loud the color of each word in Figure 9 . Since people "know" that the text color is less important than the text content, it takes a noticeable effort (and a measurable amount of extra time) to name the color of the text. Similarly, color labeling that conflicts with convention, such as labeling "stop" green, will cause confusion and errors.

Figure 9.
The Stroop Effect. See how quickly you can name the color of the text (not the color it names).

## To Quantify

Color scales are sequences of color values used to indicate relative quantity, such as those used in Figures 4 and 5. Time and time again, studies have shown that the only natural or intuitive color scales vary in lightness (value), or in saturation, often combined with a change in value (tints, tones, and shades). There are *no* perceptually-based hue scales. Yet naïve users and systems persist in trying to use hue scales for sequences of quantitative information. If this chapter can do nothing more than reduce the number of rainbow-colored visualizations like the one in Figure 5(b), it will be a success.

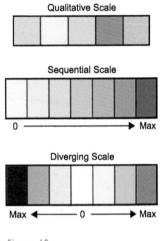

Figure 10.

Three different color scales. (From the ColorBrewer.)

Once again, we turn to cartographers to see how to use color effectively. Cynthia Brewer has created a wonderful, interactive website called "The ColorBrewer" (www.colorbrewer.org) that shows how to map color to various types of numeric information. She categorizes the color scales as *qualitative, sequential,* or *diverging.* A qualitative scale simply labels, with different hues of similar lightness or value. A sequential scale indicates quantity, varying primarily in value and/or saturation. A diverging scale is essentially two sequential scales that cross-fade through a neutral color. She recommends that this neutral be mapped to a natural mid-point in the data, such as the mean for statistical data, or zero for signed data. Figure 10 shows an example of a qualitative, a sequential, and a diverging scale, taken from the ColorBrewer. These scales have a small number of distinct levels. Quantizing continuous data into meaningful discrete bins is another important component of creating an effective color scale.

Figure 11.

Screen shot from the ColorBrewer (www.colorbrewer.org) illustrating the use of the sequential scale from Figure 10.

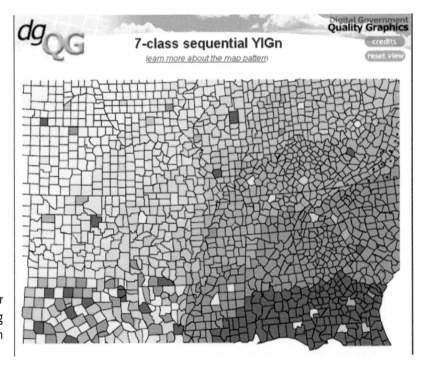

Figure 11 contains an example taken from the ColorBrewer, demonstrating the use of the sequential scale in Figure 10, where numeric values are mapped from light to dark shades of green. The data in this figure has no particular meaning, but was designed to illustrate the color scales. Figure 12 shows a diverging scale used to visualize census data, in this illustration, the change in population density. In the diverging scale, the two halves of the scales diverge from a neutral mid-point in lightness steps of contrasting colors (purple and yellow), indicating gain or loss in population.

Scales like these map a single numeric sequence, or *univariate* data, such as the population data on shown in Figure 12, or the density values in Figure 5. Other examples of univariate data include temperature, height, speed—more generally, any monotonic sequence of numeric data. *Multivariate* data involves multiple, simultaneous numeric sequences. Assigning color to multiple variables, such as population density plus change in population, results in a multivariate map as shown in Figure 13, which shows population density plus change in population. The value of multivariate

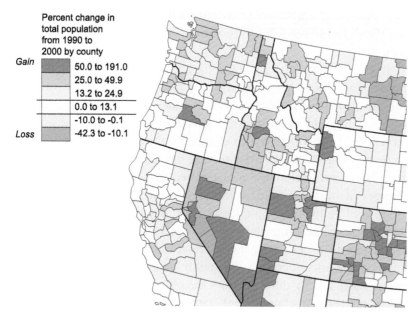

Percent change in
total population
from 1990 to
2000 by county

Gain
50.0 to 191.0
25.0 to 49.9
13.2 to 24.9
0.0 to 13.1
-10.0 to -0.1
Loss
-42.3 to -10.1

Figure 12.
This diverging color scale indicates change in population. Zero change is white. Steps away from zero are colored in increasing densities of purple (gain) or yellow (loss). (From *Mapping Census 2002*.)

Figure 13.

This multivariate scale combines change in population with population density. Population change increases color saturation; population density is indicated by hue, which also changes in value. (From *Mapping Census 2002.*)

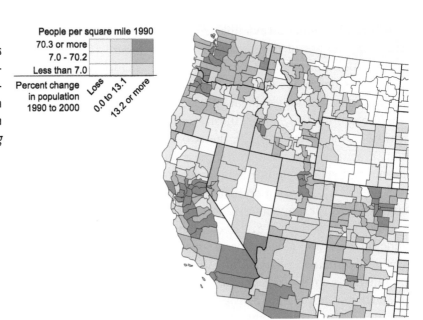

People per square mile 1990

| 70.3 or more | | | |
| 7.0 - 70.2 | | | |
| Less than 7.0 | | | |

Percent change in population 1990 to 2000

Loss   0.0 to 13.1   13.2 or more

mappings is that they allow one to more concisely display more information, and to discover relationships within the data. However, it is difficult to use color for multivariate data without causing confusion.

If a univariate color scale is a path in a color space, then multivariate data could be mapped to a plane (two variables), or to a volume (three variables), applying some rules for color mixing. For example, one could put one variable on lightness and one on saturation. Unfortunately, this example is the only 2D scheme based on perception. All other color-mixing models must be learned. Figure 13 puts saturation along one dimension—the other dimension changes in both hue and value, with green being implied as the mixture of yellow and blue (as in dyes or paints, not lights).

Estimating the ratio of colors in a mixture is difficult, even for experts. So at its best, using a color to visualize multivariate relationships only works effectively for highly quantized data, where each mixture is distinctly different from the others. In effect, it becomes a mnemonic for a labeling scheme, rather than a true analysis of mix-

Reference.

Cynthia Brewer, Trudy A. Suchan, United States Bureau of the Census. *Mapping Census 2000.* ESRI Press, 2002.

ture. Note that there are only three levels of each dimension in Figure 13, so the total mapping is only nine colors.

Another, possibly better, way to combine colored data is spatially. For example, three different sequential scales could be combined using stripes or other patterns within each region, making it easy to identify the precise quantity of each component. Any colored patterns are subject to the effects of simultaneous contrast and spatial frequency, so again, this approach is good only if the number of different colors is small.

Technologists familiar with RGB color often try to map three data values to red, green, and blue to produce a coherent visualization. For example, in the multi-spectral imaging community, different spectral bands (none of which are actually visible radiation) are mapped to each primary, then combined into a full-color image. While the resulting images can be very beautiful, I believe this approach is as fundamentally flawed for conveying information as the rainbow mapping of univariate data—there is nothing intuitive about RGB mixture, in spite of the way the cones function in the retina. The only time this approach will be helpful is if a specific numeric triple (or volume of triples around it) is of significance, and can be made to create a distinct hue to label its occurrence.

## To Indicate Shape and Size

In many instances of visualization, the goal is to represent shape, which is most easily done by shading to mimic light falling on a 3D object. Tufte calls this color as quantity also, but I find it a distinct application from the color scales discussed in the previous section. The shading to indicate hills and mountains in Figure 1 is an example of this in cartography.

The field of computer graphics converts data to perceived shape. The techniques described in Chapter 10 for realistic imagery can be applied to simulated or sampled data for visualization given a surface description of the data. For example, the rendering of a marble statue in Figure 2 in Chapter 10 used data taken from a real statue, but visualized it using computer graphics rendering.

Figure 14.

Architectural visualization. This is a rendering of a model of the staircase at one end of the Cornell University Theory Center Building. (Image by Ben Trumbore, Cornell University Program for Computer Graphics.)

Figure 14.

Architectural visualization. This is a rendering of a model of the staircase at one end of the Cornell University Theory Center Building. (Image by Ben Trumbore, Cornell University Program for Computer Graphics.)

One of the first applications of computer graphics was to design and visualize mechanical parts. Computer aided design (CAD) is now used in fields from architecture to drug design. For example, Figure 14 is an architectural rendering created at Cornell University around 1990, when such images were found only in research institutions. Now, architects routinely use such models and rendering to visualize how a building will look before it is built. The company Autodesk, for example, which sells software for 3D modeling and rendering, has a gallery of such images on its website (www.usa.autodesk.com).

Volumetric data sets create values at each point in a 3D space without any explicit surface description. Volumetric rendering is the branch of computer graphics that creates images from such data. A volumetric model can be transformed and viewed from all directions, like any 3D graphics model. Figure 15 shows tiny fragments of surgical gauze modeled by embedding them in a solid block, then alternately photographing and slicing away sections. The re-

Figure 15.

Tiny fragments of surgical gauze reconstructed from microscopic slices. (Image courtesy of Resolution Sciences Inc.)

290

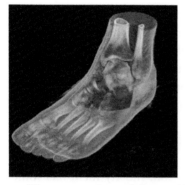

sulting data are then combined into a volumetric model and rendered to show the shape of the object.

This type of technique can be used for many forms of analysis. Often, the object is not physically cut; the slices are created non-invasively using MRI or CAT scanning technology. Figure 16 shows a reconstruction from CAT scan data of a human foot.

Figure 16.

Visualization of a human foot, reconstructed from CAT scan data. (Image courtesy of TeraRecon, Inc., rendered on the VolumePro 1000.)

The perception of shape, like the detection of edges, is caused by variation in brightness. Changing hue alone does not give an impression of shape, which can lead to some confusion about the notion of using "color" to indicate shape. More precisely, it is the shading (which is, technically, changes in color) that indicates shape. Changing hue as well, however, can add clarity by labeling, emphasizing, and mimicking nature. For example, Figure 17 is the foot in Figure 16 rendered only in shades of gray. The overall shape is still quite visible, but the colored version more clearly distinguishes the skin from the bones. In addition, the brighter hues at the silhouette edges tend to emphasize them, making the different shapes more distinct.

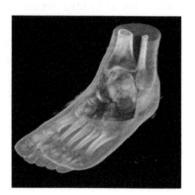

Figure 17.

The same foot as in Figure 16, shown in shades of gray. The shapes are visible, but not as clearly labeled or distinguished as in the color version.

## Making Color Robust

Color used to convey information must be robustly designed to accommodate all viewers and all environments where it is shown. This doesn't mean that all color representations must work everywhere and for everyone, but the techniques that make color robust often create designs that serve everyone more effectively.

There are two basic categories of problems that need to be solved to make color robust. The first is accommodating viewers with anomalous color vision, or colloquially, people who are "color blind." The second category is accommodating media, such as the differences between display and print. Some problems in this category can be alleviated by the correct use of color management, but if colors are far outside the target gamut, no automatic gamut-mapping algorithm will preserve the information content of those colors.

# Designing for Color Vision Deficiencies

Somewhere around 10% of the population, mostly men, has some form of anomalous color vision. To understand how best to make color robust for such individuals, we need to review some principles from Chapters 1 and 2. Color is encoded in the retina by the cones. This can be represented by the color matching functions, which capture the results of a "standard observer" matching each wavelength with some standard set of primary colors. Someone with anomalous color vision would create a different set of color matching functions than someone with normal color vision.

Color vision problems can be grouped by dimension. Anomalous trichromats see three dimensions of color, but through a significantly different set of color-matching functions (there is some variation among all observers). Dichromats see two dimensions of color, and monochromats only one. For the purposes of color information display, the variation provided by anomalous trichromats can probably be ignored, unless it is important to analyze subtle variations in hue. Monochromats are extremely rare. Most of the 10% mentioned above are dichromats.

The problems caused by dichromatic color vision are best described in terms of the opponent color model, which was presented in Chapter 2. This model, shown again in Figure 18, transforms the short (blue), medium (green) and long (red) cone responses to red-green, yellow-blue, and achromatic color channels. Dichromats fail to see color along one of the two opponent color channels.

The most common color vision problem is a genetic failure in the red-green color processing channel. People with this problem do not see redness or greenness in any color due to some failure in the

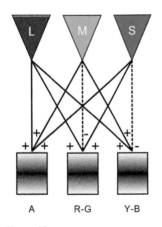

Figure 18.
The opponent color encoding.

medium or long cone processing. This makes discriminating red from green difficult, of course, but also blue from cyan, yellow from orange, or brown from gray. Monochromatic light at 500 nm, which would appear a bright, intense, green to normal observers, appears white. Intense red colors may appear black.

There are two basic forms of red-green vision problems, depending on which of the M and L cones is defective. A person with a defect in the M cone is called a *protanope*, whereas a *deuteranope* has a defective L cone. A weakness on the yellow-blue channel, caused by a defect in the S cone, is also possible, but is so rare that it is often ignored as a design constraint; such people are called *tritanopes*.

Even though dichromatic color vision seems to be caused by a failure in one of the cone photopigments, most dichromats process luminance information from all three cones. As a result, their perception of brightness is similar, if not identical, to that of persons with normal color vision. Maintaining sufficient luminance contrast is important for legibility for all viewers, and even more so for those with color vision deficiencies. As a result, the primary rule for making color robust for information visualization is the same as that for good design: Maintain sufficient value contrast so that the design can be correctly interpreted when reduced to shades of gray.

Another basic rule for making information display robust for all observers is to reinforce the color encoding with position, shape, and size information. This degree of redundancy is not always convenient, but should be provided in proportion to the importance of the information. A stop sign, for example, is not only red—it is hexagonal in shape, is located in a standard place with respect to the intersection it services, and is labeled "STOP" in white, high-contrast letters.

Brettel, Viénot, and Mollon published an algorithm in 1997 that can be used to simulate dichromatic color vision for RGB additive color. A chapter on their work is included in *Color Imaging, Vision and Technology*, edited by MacDonald and Luo. The Vischeck website (www.vischeck.com), run by Robert Dougherty and Alex Wade of Stanford University, uses these algorithms and their extensions to visualize the effect of "colorblind vision" on websites and color images.

Reference.
H. Brettel, F. Viénot, and J.D. Mollon. "Computerized Simulation of Color Appearance for Dichromats." *J. Opt. Soc. Am. A* 14 (1997), 2647–2655.

# Color and Media

Information display is traditionally designed for a specific medium. Maps and the examples in Tufte's books are created specifically for print. User interface designs are created for displays, and so on. Changing to a star-shaped model, where one design that includes colors from many sources can be output on many different media, is even more difficult than for image reproduction.

The principles of device-independent color and color management can be applied to help support cross-media information display. As in image reproduction, they make the color transformations across media more stable and predictable, but algorithms sufficient for transforming images may be less successful when applied to color for information display. For example, color scales that generate uniform steps on one medium may be transformed so that the steps become uneven, or two values become identical. Similarly, while most gamut mappings generally preserve relative lightness, they may darken or lighten extremely out-of-gamut colors, thereby reducing contrast.

Color used in information display has semantic content, which is usually expressed by color relationships. These are difficult to preserve simply by mapping each color independently as a point in some color space. An ideal system for supporting cross-media design (a topic for future research) would make it possible to explicitly specify and constrain the underlying relationships so that they could be explicitly preserved. For colors used as labels, color names could provide an appropriate level of abstraction; the name "dark red" for example, could be mapped to an appropriate color on all media, avoiding accidental transformation into the color "brown." Colors used in color scales could be represented as paths through some device-independent color space. Mapping algorithms that preserved the path relationships could then be designed. Specifying color scales as paths is already included and published in Brewer's work.

A simple example of this approach, suitable for application to charts and graphs, was implemented at Xerox PARC as part of the work on the Xerox Color Encoding Standard in the late 80s. A small set of named colors (red, green, blue, yellow, black, white, and gray),

References.

Cynthia A. Brewer. "Guidelines for Use of the Perceptual Dimensions of Color for Mapping and Visualization." *Color Hard Copy and Graphic Arts III*, edited by J. Bares, *Proceedings of the International Society for Optical Engineering (SPIE)* 2171 (1994), 54–63.

Xerox Corporation. Xerox Color Encoding Standard XNSS 288811, Xerox Corporation, 1989.

were recognized by all printer and display drivers. The mapping to device primaries was determined by the specific driver, and optimized to that device. Even black and white printers implemented these color names, choosing distinct textures or line patterns for each one so that the color encoding was still visible.

Encoding color semantics in the device driver is probably not the right implementation strategy for modern computer system architectures; however, allowing a level of indirection for color specification somewhere in the color reproduction path is a powerful way to make color used in information display more robust across media.

# A Final Example: The Colors of Emission Nebula

One of the powers of using computers to create digital images is to visualize natural phenomena that are normally impossible to see. Emission nebula are glowing clouds of gases consisting primarily of ionizing hydrogen and oxygen. The emissions may be fueled by the energy from young stars, which are created inside the nebula, or, the nebula may be the lingering results of a super nova. Seen through a telescope, they are faint, fuzzy patches in the night sky—but glowing hydrogen and oxygen should be brightly colored, as they emit nearly monochromatic red and green light, respectively.

This final section describes two different efforts to visualize the colors of emission nebula, as illustrating how art and science can be combined to make effective use of color in information display. The first example is by Thor Olson, a color engineer and avid night sky photographer. The second is the simulation created by the San Diego Super Computing Center for the Hayden Planetarium. The description here will be brief, but both images are backed by excellent web sites, which describe their creation and context.

Figure 19 is a visualization of the Veil Nebula, created by performing careful colorimetric rendering of measurements taken through a telescope using narrow band filters. The technique of using narrow filters enables a careful analysis of the spectral distribution of the light produced by the nebula at each point on the image.

Figure 19.
Colorimetrically accurate rendering of the Veil Nebula. (Image by Mike Cook and Thor Olson. For more detail, see www.nightscapes.net.)

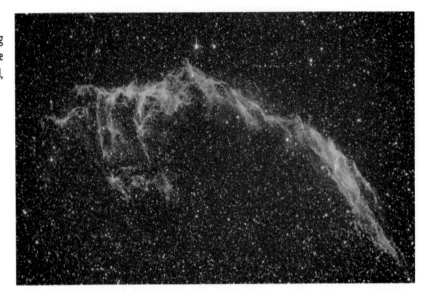

The process used by Olson to render the picture applies the principles of colorimetric color reproduction to map the spectral description to the sRGB color space. While this process was made as accurate as possible, some estimations were needed due to limitations in the data, and the process that produced the best colors for the gases created pinkish stars. As a result, the stars were color corrected separately, to make them appear white.

The resulting image shows more of the green oxygen emissions than most renderings of this phenomenon, and is potentially most enlightening for those seriously interested in the composition and hypothetical appearance of such nebula. The coloration is more subtle, however, than that often published in pictures of this nebula, which are simply photographs taken through telescopes. In such pictures, the film response, not a simulation of the human visual response, produces the colors. Other standard methods for producing color photographs of nebula involve combining several images that represent emissions in different parts of the spectrum, mapping them to red, green, and blue, and combining them to form a sort of pseudo-coloring. The result is often very colorful and attractive, and

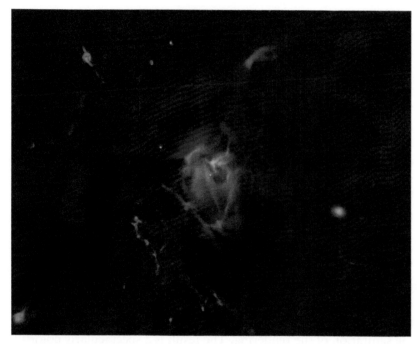

can be informative if the pseudo-color highlights significant spectral features.

The colorimetric color reproduction methods used to create Figure 19 were published at the Color Imaging Conference in 2002. Detailed information about this image, as well as other technical information about night sky photography and many lovely examples, can be found at www.nightscapes.net.

Figure 20 is a frame from an animated visualization of the birth of an emission nebula, produced by the San Diego Super Computing Center together with the American Museum of Natural History for the Hayden Planetarium. The visualization shows cosmic gases and dust condensing to form clumps, which eventually become new stars. A detailed physical simulation was used to create the time-varying volumetric model that was rendered for the visualization. As in Figure 19, the red comes from ionized hydrogen and the green from ionized oxygen, though the colors in this visualization were not colorimetrically simulated,

Reference.

Thor Olson. "The Colors of the Deep Sky."
*Tenth Color Imaging Conference*, 2002.

but were manually selected. The yellow and orange colors are simply mixtures of the red and green that appear in the same volume of space. The deep blue is entirely artificial—it represents the dust cloud, which would normally be black or dimly lit by the light of distant stars. In the early stages of the animation, there is only dust, which would make dull viewing if accurately rendered. Another enhancement was to make the stars appear larger as they became closer. Physically, they would only become brighter, as the distances are too great to see an increase in size, but this looked wrong. People's experience with moving objects at more normal distances and the fact that brighter stars make bigger dots on photographic film due to overexposure made the sequence look "more realistic" if the stars got bigger.

Figure 20 was designed to be shown in the Hayden Planetarium, which is part of the American Museum of Natural History in New York City. Seven projectors are used to produce the images on the planetarium's dome. Dome projection, however, scatters light from adjacent surfaces, which desaturates the image. To compensate for this effect, the scientists at the Super Computing Center increased the saturation of the rendering. The result is a spectacularly colorful picture, which print can only approximate. Images from the simulation, plus a more complete description of the simulation process, can be found at www.vis.ucsd.edu.

## Summary

Color in information display should follow Tufte's principles, the primary one of which is "do no harm." Color can emphasize, cluster, and label. Some combinations can indicate quantitative sequences of numbers. It is an important part of indicating shape through shading, where a smooth change in brightness can give the appearance of light falling on a 3D object. The shading need not be realistic to have its effect; the shading used by cartographers for terrain, for example, is highly stylized.

Many articles on color for information display are actually about color scales. While it has been stated many times already, let me

repeat one more time: Changing hue is a terrible way to indicate numeric sequences. There are *no* intuitive hue scales, only learned ones. There aren't even very many good learned hue scales. Thermometer scales make red "hot" and blue "cold," but black-body radiators are hotter at blue than at red. And while many learn the order of the rainbow (and hence of the spectrum) colors in grade school, which is "bigger" depends on whether you think in wavelength or frequency.

Like color design, effective use of color in information display must be learned through a combination of principles and experience. This chapter has tried to emphasize good examples, which were taken primarily from cartography. For those looking for other good examples, the National Park Service has placed their collection of maps in digital form on the web, and makes them freely available (www.nps.gov/carto/). The U. S. Census Bureau makes their publications available on the web also (www.census.gov). Like the Park Service maps, they have been put in the public domain and can be freely used.

This chapter is structured on Tufte's principles for the use of color, which have been applied by others to the problem of visualization. The "Evolution of a Numerically Modeled Storm" is a classic example of visualization created at the National Center for Supercomputing Applications. An article by Baker and Bushell describes how they used Tufte's principles to enhance their visualization of a storm system.

The emission nebula examples are included because they blend scientific simulation with the art of making effective images. Even the colorimetrically accurate simulations of human vision used by Olson had to be augmented by special treatment of the star field. The shapes and colors produced for the Hayden Planetarium were the result of a massive simulation effort, which modeled as accurately as possible a process impossible to see—the simulation covers thousands of years. The resulting images, while colored according to good scientific principles based on the light created by the simulation, also included concessions to the mass-presentation venue of a public planetarium.

Tufte's final fundamental use of color is to enliven or decorate.

Reference.

M.P. Baker and C. Bushell. "After the Storm: Considerations for Information Visualization." *IEEE Computer Graphics and Applications* (1995).

Good illustrations are beautiful, which is why people collect maps and make posters of the route map of the London Underground. The value of aesthetics is often questioned in "practical" disciplines like science and engineering, however, some of the principles that make color design more attractive, such using a limited palette (eliminates confusion), with a range of tints and tones (to provide luminance contrast), make color more effective as well. An attractive illustration will be looked at more carefully and remembered longer, which is, after all, the primary goal of information display.

# Bibliography

This bibliography contains a selected set of reference books for the different fields of digital color. Rather than simply providing a list, this section provides a brief discussion about these books, grouped by topic. These are primarily books that are in my personal library, so both the selection and the description are inevitably biased, and have not all been studied in equal detail. However, a primary goal of this book is to be a guide, and I felt that a discussion, even if biased, would be more helpful than a simple, alphabetical list.

Many of these books are described within the chapters. Those recommendations will be repeated here as well—the goal is to make this section stand alone as resource for finding the right "next book" after having read this one.

## Color Vision and Measurement

There are many excellent color vision books. Brian Wandell's *Foundation of Vision* [50] is both fairly recent and has a computational flavor that reflects Brian's interest and expertise in digital imaging.

The text by Leo Hurvich, *Color Vision* [24], is out-of-print but may be available used or in academic libraries or used. It explains color vision in terms of the opponent color model, and includes a detailed analysis of color vision deficiencies based on that model. Cornsweet's 1970 book, *Visual Perception* [6] is also a classic, though now 30 years old.

Mark Fairchild's book, *Color Appearance Models* [10] is an excellent reference on color appearance models as being pursued by the color imaging community. It describes the basic concepts and principles in this field, and also provides a snapshot of the computational models being developed at the time (1998). The book has recently gone out of print, but Mark is planning a second edition for early in 2004.

*Interaction of Color* by Josef Albers [2] is a sequence of exercises that help the student discover the relative nature of color, all to be performed with colored paper. It includes some wonderful examples of how spatial relationships affect color appearance. The original publication was a large, expensive book with many color plates, which is now out-of-print. There is, however, a pocket edition with the same text but fewer plates that is well worth its modest cost.

Wyszechki and Stiles' massive *Color Science* [57] is the standard reference for colorimetry and color psychophysics. A substantial number of pages of this book contain tables of numbers, reflecting the importance of printing such information pre-web and pre-CDROM. It is interesting to note that this book has no color pictures.

Another classic book on color science and its application is Judd and Wyszechki's *Color in Business, Science, and Industry* [28]. A more recent reference on color measurement is Hunt's *Measuring Color* [23].

*The Science of Color* [44] is a collection of recent articles edited by Steven Shevell that are primarily about color vision, although they also cover topics such as digital color technology and the physics of color. It is not yet released, but I recognize many of the contrtibutors as authorities in their fields.

Roy Bern's update of the classic Billmeyer and Saltzman *Principles of Color Technology* [5] is an up-to-date reference on color, its perception, and its measurement. It includes an excellent discussion of the evolution of color difference spaces, such as CIE94, as well as detail about other perceptually-based color order systems beyond those described in this book.

# Color in Nature

The *Color of Nature* [37] is a beautiful book created by the staff of the Exploratorium, a hands-on science museum in San Francisco. While primarily a picture book with brief explanations, the examples and illustrations are quite wonderful. The Exploratorium website (www.exploratorium.edu) is also a useful resource.

*Light and Color in Nature and Art* by Williamson and Cummins, [54] was designed for an undergraduate course. It is an especially useful reference for the physical and optical principles of light, black-body radiation, diffraction, interference, pigments and paints. While it claims to be for a general undergraduate audience, it contains plenty of equations and technical detail.

Lynch and Livingston are astronomers with a devotion to photographing and explaining the different color phenomena seen in the sky. Their *Color and Light in Nature* [32] is full of technical detail and wonderful pictures.

Philip Ball's *The Bright Earth* [3] describes how the evolution of colored pigments has influenced art throughout the ages. Ball is a chemist, but the book does not require a technical background to appreciate it—in fact, it has become somewhat of a commercial best seller. Also fascinating from a historical perspective is *Mauve* [14], the biography of William Perkin, who invented and commercialized the first coal tar dye. Without these synthetic dyes, the industrialized world would be a much less colorful place.

I have had recommended, but do not personally own, *The Physics and Chemistry of Color* by Kurt Nassau [38]. It is said to be a highly detailed and technical reference.

# Color Reproduction

The three color reproduction industries are printing, photography, and television. Of these three, I have been most involved in printing, which is reflected in my library.

Hunt's *The Reproduction of Colour In Photography, Printing & Television* [22] spans all three of the major color reproduction industries, and is simply invaluable as a reference for color reproduction. The fifth edition was published in 1996. Dr. Hunt is still active in the digital imaging community, and is a regular speaker at the SID/IS&T Color Imaging Conference.

The classic book on color printing is John Yule's *Principles of Color Reproduction* [59], which has been out of print for many years. A new edition, co-authored with Gary Field [58], was released in 2001. This is essentially the original Yule book, with an additional section appended that discusses how the graphic arts have changed since the original edition was published in 1967.

Field has written his own book, *Color and Its Reproduction* [11]. It is both detailed and beautifully presented, and is a much better reference for modern printing practice than Yule.

Phil Green's *Understanding Digital Color* [17] is an excellent reference for the all-digital world of the modern graphic arts, including digital scanning, printing and color management. It is also very nicely illustrated.

Charles Poynton has written two excellent books on video technology: *A Technical Introduction to Digital Video* [39] and *Digital Video and HDTV Algorithms and Interfaces* [40]. Charles' books are also an outstanding reference for the details and subtleties of digital image encoding.

The *Focal Encyclopedia of Photography* [45] is a general reference for the technology of photography, and more broadly, imaging. Written as an encyclopedia, with contributions from many experts, it provides a concise, authoritative summary of topics such as densitometry and sensitometry. The same authors, Leslie Stroebel and Richard Zakia, have written *Basic Photographic Materials and Processes* together with John Compton and Ira Current [46]. This book is a technical overview of photography, written for the serious student of photography.

*Projection Displays,* by Stupp and Brennesholtz [48], is a complete and authoritative reference for the technology of projectors and projection systems. Published in 1999, it doesn't include the absolute latest advances in digital projectors (which are evolving at a great rate), but it should be on the shelf of anyone seriously working with projection displays.

Lindsay MacDonald is a prolific author and editor. The book he edited with Anthony Lowe, *Display Systems* [33], is a comprehensive survey of different display technologies. He has co-edited with Ronnie Luo two collections of articles that provide an excellent overview of current research in digital color and digital imaging: *Colour Imaging Vision and Technology* [35], published in 1999, and *Colour Image Science: Exploiting Digital Media* [36], published in 2002. The articles are written as summaries of specific research domains, and as such, provide both an introduction and a description of the current state-of-the-art in that area.

Gaurav Sharma's *Digital Color Imaging Handbook* [42] is a similar collection of articles, but more focused on color imaging technology.

The *Encyclopedia of Imaging Science and Technology* [21] is a large, very expensive, two-volume set of articles on color in a wide range of fields. I cannot comment on the general quality of the contents as I do not own it, but it does include an excellent article on display characterization by David Brainard. A draft version of this article, as well as other useful references, ca be found on David's website (color.psych.upenn.edu/brainard/).

Giorgianni and Madden's book, *Digital Color Management: Encoding Solutions* [15], published in 1998, was the first comprehensive book on this topic. It is directed toward the engineers and programmers who implement digital color technology and systems. It has a beautifully concise introduction to basic color perception and colorimetry, which has influenced my presentation of those concepts. The book presents a detailed discussion of cross-media color transformations and the various tradeoffs that must be considered when implementing them.

Phil Green and Lindsay MacDonald have edited *Colour Engineering: Achieving Device Independent Colour* [18], a collection of articles that provides a more recent (published in 2002) reference for implementing the components of device-independent color management systems.

*The GATF Practical Guide to Color Management* [1], as the title suggests, is a guide for those interested in using color management, as opposed to implementing it. Aimed at the professional graphic arts industry, it includes product descriptions as well as general practices. *Understanding Digital Color* [17] is also a good reference for the practical application of color management.

It is almost impossible to practice digital color imaging without using Adobe System's Photoshop. There are dozens of trade books on Photoshop and its application. The classic use of Photoshop is for prepress. Dan Margulis is a well-respected expert in this field. His *Professional Photoshop: The Classic Guide to Color Correction* [36] is designed for the prepress professional, whose goal is making good color separations specified as CMYK images. Margulis is quite scathing about the limitations of device-independent color in this context. *Adobe Photoshop 6.0 for Photographers,* by Martin Evening [8], reflects a growing field in digital imaging and the use of tools such as Photoshop. Many digital printers accept RGB input, hiding the conversion to CMYK from the user. Therefore, the user can focus on color management strictly in terms of RGB color spaces. Evening's book contains a nice description of RGB color spaces for the lay person, including the factors that affect the choice of a suitable RGB working space. There is a more recent version for Photoshop 7.0 [9].

Digital color documents are often distributed as PDF or PostScript, the document description formats created by Adobe. Their common imaging model, which is described in the *PostScript Language Reference Manual*, includes complete support for device-independent color.

# Computer Graphics

Compared to the color reproduction industries, computer graphics is a relatively young field. The first professional conferences on the topic were in the mid-70s, and the first text books came out at the end of that decade. The book that became the most popular text book in the field, *Fundamentals of Interactive Graphics*, by Jim Foley and Andy van Dam [12], was first published in 1982. The second edition, *Computer Graphics Principles and Practice* [13], published in 1990, added authors Steve Feiner and John Hughes and substantially increased the size and scope of the book. The field of computer graphics has expanded significantly since then, making it difficult to create a single, comprehensive text.

Roy Hall's *Illumination and Color in Computer Generated Imagery* [20] focuses on light and color as it was applied in graphics in the late 80s. It is still one of the most referenced books on color in graphics, although it has been out-of-print for many years.

Andrew Glassner's *Principles of Digital Image Synthesis* [16] is a large, two-volume text on all aspects of image synthesis in computer graphics, including color. It even has a detailed description of the Kubelka-Munk model for simulating paint color. Andrew has written other books on graphics and rendering techniques as well, and writes a regular column in the technical magazine, *Computer Graphics & Applications*.

Dave Roger's books on graphics are clear, concise references, with a strong emphasis on the mathematical and algorithmic aspects of graphics. The second edition of *Procedural Elements for Computer Graphics* [41] contains a good chapter on color, including device-independent color and printing, and transformations between RGB color spaces.

Henrik Wann Jensen's *Realistic Image Synthesis Using Photon Mapping* [27] includes a good introduction to the modern global illumination models that make modern computer graphics images rival those taken with a camera.

There are two recent books on non-photorealistic rendering: *Non-photorealistic Rendering,* by Amy and Bruce Gooch [19], and *Non-Photorealistic Computer Graphics: Modeling, Rendering and Animation,* by Thomas Strothotte and Stefan Schlechtweg [47]. Both provide a comprehensive overview of the techniques used to make computer graphics images look like paintings and illustrations.

There are far more users of computer graphics systems than implementers. Two of the most widely used developer's platforms are RenderMan and OpenGL. Both are available online, and have excellent references written for them: *The RenderMan Companion* by Steve Upstill [50]and *OpenGL Programming Guide, Third Edition,* by Mason Woo, Jackie Neider, Tom Davis and Dave Shreiner [56]. Microsoft's native 3D platform for Windows is called DirectX (previously Direct3D). Books on this topic seem directed primarily at game developers. The native 3D platform for the Macintosh system is OpenGL. POV-Ray is an open source ray tracing package that is widely used. It is available for free download at www.povray.org.

# Color for Design and Information Display

Wucius Wong's books on design are wonderfully concise and beautifully illustrated. His *Principles of Color Design* [55] is my favorite reference for color and design. His methodology is based on the Munsell system, and is very orderly and algorithmic.

Johannes Itten is considered by many to be an important authority on art and color, but I am not personally familiar with his books. *The Art of Color* [25] is frequently referenced.

Most books on color and design are primarily sample sets of swatches and color schemes. Pantone is famous for their palettes of ink colors, which can either be printed as custom inks or as specified blends of CMYK colors. The *Pantone Guide to Communicating with Color* [7] is a color scheme book based on the Pantone inks. *Color Harmony 2* by Bride Whelen [53] is another example, which also includes a CDROM with an interactive palette tool called "The Palette Picker." This tool used to be available for download also, but hasn't been maintained or upgraded since its release in 1994.

Berlin and Kay's *Basic Color Terms* [4], originally published in 1969, established the universal nature of color names across language and cultures. Further study, which is well summarized in Lammen's thesis [30], establishes that there is something fundamental about colors and names for basic colors such as red, green, orange, and yellow.

Commercial use of color names is widespread. However, Kelly and Judd's *Color, Universal Language and Dictionary of Names* [29] published in 1976, is the only formal effort I know of to establish standards for color naming beyond the basic colors.

Edward Tufte's books are essential references on design for information display, but only *Envisioning Information* [49] discusses color. The basic principles presented there should guide all applications of color in information display.

Colin Ware's *Information Visualization* [52] provides a perceptual foundation for many of the practices of information display. It is written from the perspective of research and application of computer-aided visualization, rather than from a strictly visual design standpoint like Tufte.

Haim Levkowitz's *Color Theory and Modeling for Computer Graphics, Visualization, and Multimedia Applications* [31] summarizes many interesting aspects of color as applied to graphics and visualization, but doesn't supply the depth that Ware's book does. It is also oddly produced, as it contains only black/white images for all of the color illustrations, directing the reader to the full-color versions on the web.

Anne Spalter's *The Computer in the Visual Arts* [43] is written primarily to present digital technology to the visual artist. However, it also provides insight to the technologist on the visual/artistic implications of digital color media. She includes a detailed description of digital color selection tools, providing one of the few easily acquired references for this topic.

*Computer Generated Color* by Richard Jackson, Lindsay MacDonald and Ken Freeman [26] is in many ways the predecessor to this book. Also full color throughout (which is still rare for technical books on color), it summarizes many important factors in the use of digital color for presentation and display. It includes more details on display technology, and has less emphasis on device-independent color than this book, as color management was not common in 1994. While showing its age in places, it is still a useful summary and guide.

# The Books

[1] Richard M. Adams II and Joshua B. Weisberg. *The GATF Practical Guide to Color Management.* Pittsburgh: GATFPress, 2000.

[2] Josef Albers. *Interaction of Color.* New Haven: Yale University Press,1963.

[3] Philip Ball. *Bright Earth: Art and the Invention of Color.* New York: Farrar, Straus and Giroux, 2002.

[4] Brent Berlin and Paul Kay. *Basic Color Terms: Their Universality and Evolution.* Berkeley, CA: University of California Press, 1991. Originally published in 1969.

[5] Roy S. Berns. *Billmeyer and Saltzman's Principles of Color Technology, Third Edition.* New York: John Wiley & Sons, 2000.

[6] Tom N. Cornsweet. *Visual Perception.* New York: Academic Press, 1970.

[7] Leatrice Eiseman. *Pantone Guide to Communicating with Color.* Sarasota, FL: Design Books International, 2000.

[8] Martin Evening. *Adobe Photoshop 6.0 for Photographers.* Boston, MA: Focal Press, 2001.

[9] Martin Evening. *Adobe Photoshop 7.0 for Photographers.* Boston, MA: Focal Press, 2002.

[10] Mark D. Fairchild. *Color Appearance Models.* Reading, MA: Addison-Wesley, 1998.

[11] Gary G. Field. *Color and Its Reproduction, Second Edition.* Pittsburgh: GATFPress, 1999.

[12] James D. Foley and Andries Van Dam. *Fundamentals of Interactive Computer Graphics.* Reading, MA: Addison-Wesley, 1982.

[13] James D. Foley, Andries Van Dam, Steven K. Feiner, and John F. Hughes. *Computer Graphics: Principles and Practice, Second Edition.* Reading, MA: Addison-Wesley, 1990.

[14] Simon Garfield. *Mauve: How One Man Invented a Color That Changed the World.* New York: W.W. Norton & Company, 2002.

[15] Edward J. Giorgianni and Thomas E. Madden. *Digital Color Management: Encoding Solutions.* Reading, MA: Addison-Wesley, 1998.

[16] Andrew S. Glassner. *Principles of Digital Image Synthesis.* San Francisco: Morgan Kaufmann, 1995.

[17] Phil Green. *Understanding Digital Color, Second Edition.* Pittsburgh: GATFPress, 1999.

[18] Phil Green and Lindsay MacDonald (Editors). *Colour Engineering: Achieving Device Independent Colour.* New York: John Wiley & Sons, 2002.

[19] Amy Gooch and Bruce Gooch. *Non-Photorealistic Rendering*. Natick, MA: A K Peters, 2001.

[20] Roy Hall. *Illumination and Color in Computer Generated Imagery*. New York: Springer-Verlag, 1989.

[21] Joseph P. Hornak (Editor). *The Encyclopedia of Imaging Science and Technology*. New York: John Wiley & Sons, 2001.

[22] Robert W.G. Hunt. *The Reproduction of Colour In Photography, Printing & Television, Fifth Edition*. Tolworth, England: Fountain Press, 1996.

[23] Robert W.G. Hunt. *Measuring Colour, Third Edition*. Tolworth, England: Fountain Press, 1998.

[24] Leo M. Hurvich. *Color Vision*. Sunderland, MA: Sinauer Associates Inc., 1980.

[25] Johannes Itten. *The Art of Color: The Subjective Experience and Objective Rationale of Color*. New York: John Wiley & Sons, 1997. (Revised edition.)

[26] Richard Jackson, Lindsay MacDonald, and Ken Freeman. *Computer Generated Colour, A Practical Guide to Presentation and Display*. New York: John Wiley & Sons, 1994.

[27] Henrik Wann Jensen. *Realistic Image Synthesis Using Photon Mapping*. Natick, MA: A K Peters, 2001.

[28] Deane B. Judd and Günter Wyszecki. *Color in Business, Science, and Industry*. New York: John Wiley & Sons, 1975.

[29] Kenneth L. Kelly and Deane B. Judd. *Color, Universal Language and Dictionary of Names*. Washington, DC: National Bureau of Standards (U.S.) Special Publication 440, 1976.

[30] Johan M. Lammens. "A Computational Model of Color Perception and Color Naming," Ph.D. diss., State University of New York at Buffalo, 1994. Also available on CiteSeer (citeseer.nj.nec.com/cs).

[31] Haim Levkowitz. *Color Theory and Modeling for Computer Graphics, Visualization, and Multimedia Applications*. Boston: Kluwer Academic Publishers, 1997.

[32] David K. Lynch and William Livingston. *Color and Light in Nature, Second Edition*. Cambridge, UK: Cambridge University Press, 2001.

[33] Lindsay W. MacDonald and Anthony C. Lowe (Editors). *Display Systems*. New York: John Wiley & Sons, 1997.

[34] Lindsay W. MacDonald and M. Ronnier Luo (Editors). *Colour Imaging: Vision and Technology*. New York: John Wiley & Sons, 1999.

[35] Lindsay W. MacDonald and M. Ronnier Luo (Editors). *Colour Image Science: Exploiting Digital Media*. New York: John Wiley & Sons, 2002.

[36] Dan Margulis. *Professional Photoshop: The Classic Guide to Color Correction*. New York: John Wiley & Sons, 2001.

[37] Pat Murphy and Paul Doherty. *The Color of Nature*. San Francisco: Chronicle Books, 1996.

[38] Kurt Nassau. *The Physics and Chemistry of Color: The Fifteen Causes of Color, Second Edition*. New York: John Wiley & Sons, 2001.

[39] Charles A. Poynton. *A Technical Introduction to Digital Video*. New York: John Wiley & Sons, 1996.

[40] Charles A. Poynton. *Digital Video and HDTV Algorithms and Interfaces*. San Francisco: Morgan Kaufmann, 2002.

[41] David F. Rogers. *Procedural Elements for Computer Graphics, Second Edition*. Boston: McGraw-Hill, 1998.

[42] Gaurav Sharma. *Digital Color Imaging Handbook*. Boca Raton, FL: CRC Press, 2002.

[43] Anne Morgan Spalter. *The Computer in the Visual Arts*. Reading, MA: Addison-Wesley, 1999.

[44] Steven K. Shevell. *The Science of Color, Second Edition*. London: Elsevier Science Ltd., 2002.

[45] Leslie Stroebel and Richard Zakia (Editors). *Focal Encyclopedia of Photography, Third Edition*. Boston: Focal Press, 1993.

[46] Leslie D. Stroebel (Editor), John Compton, Ira Current, Richard D. Zakia. *Basic Photographic Materials and Processes*. Boston: Focal Press, 1990.

[47] Thomas Strothotte and Stefan Schlechtweg. *Non-Photorealistic Computer Graphics: Modeling, Rendering and Animation*. San Francisco: Morgan Kaufmann Publishers, 2002.

[48] Edward H. Stupp and Matthew S. Brennesholtz. *Projection Displays*. New York: John Wiley & Sons, 1999.

[49] Edward R. Tufte. *Envisioning Information*. Cheshire: Graphics Press, 1990.

[50] Steve Upstill. *The RenderMan Companion: A Programmer's Guide to Realistic Computer Graphics*. Reading, MA: Addison-Wesley, 1989.

[51] Brian A. Wandell. *Foundations of Vision: Behavior, Neuroscience and Computation*. Sunderland, MA: Sinauer Associates Inc., 1995.

[52] Colin Ware. *Information Visualization: Perception for Design*. San Francisco: Morgan Kaufmann, 2000.

[53] Bride M. Whelan. *Color Harmony 2*. Rockport, MA: Rockport Publishers, 1994.

[54] Samuel J. Williamson and Herman Z. Cummins. *Light and Color in Nature and Art*. New York: John Wiley & Sons, 1983.

[55] Wucius Wong. *Principles of Color Design, Second Edition*. New York: John Wiley & Sons, 1997.

[56] Mason Woo, Jackie Neider, Tom Davis, and Dave Shreiner. *OpenGL® Programming Guide: The Official Guide to Learning OpenGL, Version 1.2, Third Edition*. Reading, MA: Addison-Wesley, 1999.

[57] Günter Wyszecki and W.S. Stiles. *Color Science, Second Edition*. New York: John Wiley & Sons, 1982.

[58] John A.C. Yule and Gary G. Field. *Principles of Color Reproduction*. Pittsburgh: GATFPress, 2001.

[59] John A.C. Yule. *Principles of Color Reproduction*. New York: JohnWiley & Sons, 1967.

# Journals

Most journals are associated with the professional societies described in the next section. The exception is *Color Research and Application*, published by John Wiley & Sons. It reports on the "science, technology, and application of color in business, art, design, education, and industry," and is one of the primary publications of the color science community.

# Professional Societies

Below are many of the professional societies in the fields of digital color, along with their websites and primary journals. Some of the larger groups sponsor more conferences and journals than mentioned here; those listed are most likely to be of interest to those involved in digital color. As in the bibliography, the list is biased by my experience and nationality.

### AIC Color (originally Association Internationale de la Couleur)

Website: www.aic-color.org

AIC Color is an international, interdisciplinary organization of color organizations. The AIC's charter reads: "To encourage research in all aspects of color, disseminate the knowledge gained from this research, and promote its application to the solution of problems in the fields of science, art, design, and industry on an international basis."

## The Inter-Society Color Council (ISCC)

Website: www.iscc.org
The ISCC claims to be the principal professional society in the field of color in the United States, encompassing the arts, sciences, and industry. Its membership includes primarily other color organizations and societies; the number and the diversity of their membership is a strong reminder of how many fields study and apply color.

## The Society for Imaging Science and Technology (IS&T)

Website: www.imaging.org
**Journals:** *Journal of Electronic Imaging*
*Journal of Imaging Science and Technology*
The Society for Imaging Science and Technology is an international non-profit organization whose goal is to keep members aware of the latest scientific and technological developments in the field of imaging. It focuses on imaging in all its aspects, with particular emphasis on digital printing and electronic imaging. It sponsors an impressive list of conferences in all fields of digital imaging and digital color. Of note are the SID/IS&T Color Imaging Conference, and the Electronic Imaging Conference.

## The Optical Society of America (OSA)

Website: www.osa.org
**Journal:** *Journal of the Optical Society of America A (JOSA A)*
The mission of OSA is to "promote the generation, application and archiving of knowledge in optics and photonics," which includes information about color and color vision. OSA hosts conferences and workshops on a wide range of topics in optics and photonics. Many color vision professionals publish in its journals and at its conferences.

## The Society for Information Display (SID)

Website: www.sid.org
**Journal:** *Journal of the Society for Information Display*
**Magazine:** *Information Display*
The Society for Information Display includes professionals in all of the technical and business disciplines that relate to display research, design, manufacturing, applications, marketing, and sales. It is the primary technical organization for those interested in display technology of all forms.

Its annual conference is the premiere publication venue for advances in display technology.

### Association for Research in Vision and Ophthalmology, Inc. (ARVO)

Website: www.arvo.org
**Journal:** *Journal of Visual Science*
From their website: "The purposes of ARVO shall be to encourage and assist research, training, publication, and dissemination of knowledge in vision and ophthalmology." The ARVO conference and journal is one of the publishing venues for those doing research in color vision.

### The Society of Motion Picture and Television Engineers (SMPTE)

Website: www.smpte.org
**Journal:** *SMPTE Journal*
SMPTE is the leading technical society for the motion imaging industry, which includes both film and television. SMPTE publishes ANSI-approved standards, recommended practices, and engineering guidelines for the television and movie industries. It hosts both research and industry-oriented conferences.

### Technical Association of the Graphic Arts (TAGA)

Website: www.taga.org
**Journal:** *Journal of Graphic Technology*
TAGA is a global technical organization for the graphic arts. TAGA focuses on graphic arts systems, software, and computer technology developments, as well as the more traditional areas of press, ink, and paper engineering applications. The annual TAGA conference contains papers both from research and from practice.

### The Graphic Arts Technical Foundation (GATF)
### Printing Industries of America (PIA)

Website: www.gain.org
PIA/GATF combine two industrial organizations for the graphic arts and printing. Their information portal is called GAIN (Graphic Arts Information Network). The GAIN website offers access to a wide variety of services and publications. GATF establishes standards and standard practices for the prepress industry. For example, the SWOP recommendations for printing ink composition and coverage, which are widely used

in printing in the United States, were created by GATF. GATFPress publishes a wide range of technical books and handbooks.

## Association of Computing Machines (ACM)

Website: www.acm.org
Journal: *Communications of the ACM*
The Association of Computing Machines (ACM) is the professional organization for computer science professionals. ACM sponsors Special Interest Groups (SIGs) on a variety of topics, many of which may include some aspect of digital color. The largest of these is graphics (SIGGRAPH), which is described more fully below. Other SIGs involved in digital color include human-computer interaction (SIGCHI), Hypertext, Hypermedia and Web (SIGWEB), and Multimedia (SIGMM).

## ACM SIGGRAPH

Website: www.siggraph.org
Journals: *ACM Transactions on Graphics*
        *(SIGGRAPH Annual Conference Proceedings*
SIGGRAPH is the primary professional organization for those interested in computer graphics, and the SIGGRAPH annual conference is considered by many in the field to be the premiere publishing venue for research papers in graphics, to the extent that its proceedings carry the same weight as a journal. Since 2001, the conference proceedings have been published special issue of *Transactions on Graphics*. As well as its annual conference, SIGGRAPH co-sponsors smaller conferences and workshops on the topic of graphics and interaction techniques.

## IEEE Computer Society

Website: www.computer.org
Journal: *IEEE Transactions on Visualization and Computer Graphics*
Magazines: *Computer Graphics & Applications; Multimedia*
The IEEE is the Institute of Electrical and Electronics Engineers, a professional organization founded for those involved in electrical and electronic engineering. The Computer Society is the branch of IEEE focused on computing. The publications and conferences of the IEEE Computer Society complement those of ACM. The annual IEEE Visualization conference and the associated IEEE Symposium on Information are the primary conferences on computer-aided visualization.

# Standards

### International Color Consortium (ICC)

Website: www.color.org

The ICC is an industry consortium, established in 1993 "for the purpose of creating, promoting, and encouraging the standardization and evolution of an open, vendor-neutral, cross-platform color management system architecture and components." This is the organization that establishes the standard specification for the ICC profiles used in color management systems.

### International Commission on Illumination (CIE)

Website: www.cie.co.at/cie/

CIE Image Technology Website: www.colour.org

The CIE, from its French title, Commission Internationale de l'Eclairage, is an international organization known primarily for its recommendations and standards, such as those that underlie CIE colorimetry. Its membership is composed of national committees.

The topics of interest to the CIE are organized into divisions. Division 8, Image Technology, is the one most concerned with defining standards for digital color. Its committees include the ones on Color Appearance Models (CIECAM97, CIE2002), Gamut Mapping, Communicating Color Information, etc.

### The International Organization for Standardization (ISO)
### International Electrotechnical Commission (IEC)

Websites: www.iso.ch; www.iec.ch

ISO is an international organization of standards bodies, founded to promote international standards for exchange in science and industrial applications. Similar to the CIE Divisions, ISO standards are created by from a hierarchy of technical committees, subcommittees, and working groups.

ISO has the charter to manage standards for all "technical fields" except electrical and electronic engineering standards, which are controlled by the IEC. The two organizations also develop standards jointly. Color spaces, compression, and other imaging standards, such as those that form the basis for sRGB, JPEG, and MPEG, are ISO/IEC standards.

# Index

## Symbols

3 x 3 matrix 47–48, 54, 59, 148, 158, 198
density 173
3D graphics 223
8 bit encoding 243
8-bit signals 146

## A

absolute colorimetric 204
absorption 80
   dye layers 171
   transmitted or reflected light 171
adaptation 34
   brightness 120
additive
   color 91
   color matching experiments 10
   color reproduction 147
   color systems 137, 137–160, 156
   output system 147
afterimages 28
alpha 235
alpha channel 146
alpha transparency 235
analogy 257
animal vision 7
anology 278
antialiasing 236
Apple ColorSync 217
Autodesk 290
Automatic White Balance (AWB) 121

## B

background 37
Barco Calibrator 149
Bayer array 120
black
   adding 170
   for contrast in printing 164
   ink for contrast 183
black-body curve 14
black-body radiators 67, 83, 228
block filters 166
brightness 3, 28, 36, 43–62, 140
   additive color systems 157
   maping digital inputs 140
   measurement 29
   non-uniform encoding 50
   of color 3
   scale 48–49
   standard encoding 50
   tone mapping 95
brightness scales 44
   linear reflectance 50
   perceptually uniform 49
Brown Animation Generation System 230
bubble jet printer 185
bump mapping 233

## C

CAD 290
calibration 209
camera systems 119
Canon Color Advisor 264
Cathode Ray Tubes (CRT) 138
CCD array 119
CFA 120

characterizing 54, 92
   additive color systems 156
   cameras 131

color technology 89
commercial systems 157
DLP projectors 159
film recorders 188
image capture systems 130
images 133
RGB 54
software 157
subtractive color systems 187, 188
chroma 36, 127
chromatic aberration 261
chromatic adaptation 34, 35, 120
chromaticity diagram 13
   additive color gamuts 139
   CRT display 55
   plotting RGB 47
CIE 12
CIE Colorimetry 12
CIE tristimulus values 12
CIE94 27
CIECAM02 40
CIELAB 25, 197, 261, 264, 268
   color management 201
   scanners 131
CIELUV 25, 270
CMM 197, 217
CMOS arrays 119
CMY 164
   filters 166
   neutral gray 169
CMYK 107, 268
   adding black 169
   color separations 179
   digital image encoding 124
   imagesetters 184
color
   as beauty 278
   as information 277–300
   as representation 278
   on color 261

319

Printed and bound by CPI Group (UK) Ltd, Croydon, CR0 4YY

23/10/2024

01777682-0001